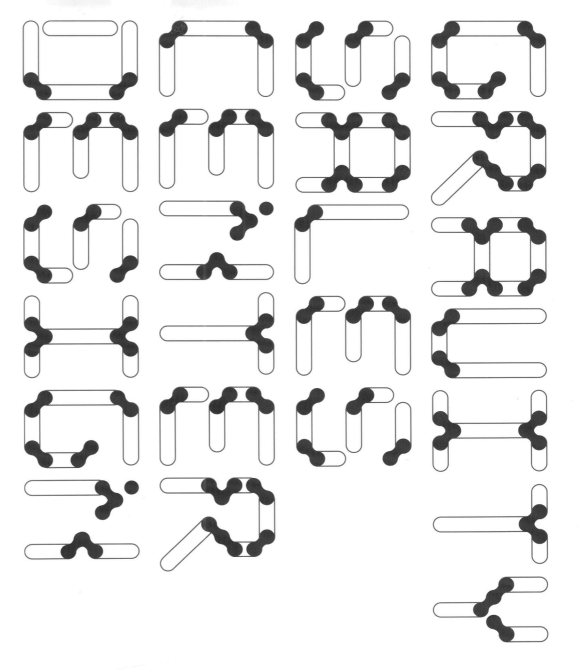

万有引力

售楼部设计 XI

GRAVITY
/
SALES CENTER
DESIGN XI

欧朋文化 策划　　黄滢 马勇 主编

华中科技大学出版社
http://www.hustp.com
中国·武汉

PREFACE 序言1

临时建筑的有限与无限可能
——关于接待中心的本质性思考

域研近相空间设计
李俊平 / 康智凯

域研近相空间设计主持建筑师
华梵大学建筑系现任讲师
2010年台湾室内设计大奖临时建筑类TID奖
2012年IAI亚太设计双年奖样板房类优秀奖
2012年台北设计奖公共空间类铜奖

临时性建筑，泛指临时使用的建筑物或构筑物，通常是指在使用目的上有一定的时间限制，在法令的规范上也有一定的使用期限，在期限届满后应当拆除的建筑物或构筑物。以往临时性建筑以展览性质为主，较为著名的如伦敦蛇形艺廊（Serpentine Gallery），每年夏天都会邀请著名的建筑师于附近的草地上打造寿命只有三个月的临时构筑物，在不考虑实用性的条件下充分赋予建筑师创作上的无限自由，尽情阐述自己的建筑理念。现在的蛇形艺廊已经成为国际建筑大师们实现自己建筑理念的最好建筑实验场，同时也在每年展期期间为伦敦肯辛顿花园（Kensington Gardens）带来了超过75万人次来访的巨大观光效益。然而，近二十几年来，以接待中心与样板房为主的临时性建筑，作为房地产销售的主要道具已成为极具台湾特色的特殊商业模式。虽然，这种遍地开花的模式一直存在着许多争议，但不可讳言的是，这已经成为房产销售必备的消费模式，甚至是一种文化，一种独特的地景式空间的商品化过程。

接待中心是在当时的时空背景下，为了鼓励建设公司在资金不够充裕的同时还能进行土地建筑开发的一个权宜之计——"房屋预售制度"。整个制度经过数十年来的演变及人们生活素质的提升，消费者已从早期的实用性质提升到空间体验的层次，设计师也在代销公司的推波助澜下不断地拿出创新与前卫的设计理念，在这个领域里找到了设计能量的出口。

从宏观的角度来说，接待中心就像是都市地景里的昙花一现，总能在看似规矩有序的地景里偶然绽放出不受拘束的光彩与姿态。或顺应都市纹理，或摆脱规矩框架，除了实际的使用功能外，从某种都市的意义层面来看，接待中心存在的价值，反而像是件大型的地景艺术，以雕塑般的形态为地景增添活力，为城市增添光彩。一个像台北这样的城市，满满的能量与生命力俯首可见，但却总缺乏一些美学上的惊喜与幽默，接待中心，刚好是一个机会点，一种美学思考的介入，一种公共地景艺术的机会介入，让这种昙花一现式的瞬间，不会永远只有浓浓的商业气息，而是至少能以一种都市创意发想的姿态，提醒这个城市艺术气息的魅力与重要性。

既然接待中心本质是临时性建筑物，那就更应该做到一些永久性建筑无法做到的事。对于建筑师而言，在层层法规的规范下所填满的令人窒息的公共建筑环境中，接待中心的出现，像是建筑废墟中的一个绿洲，在法规的灰色暧昧地带中如海市蜃楼般地尽情奔放。永久性建筑，特别是在房产销售的领域，受限于法规等诸多因素的限制，建筑师通常在有限的条件下为业主做完合理规划与最大的坪效后往往已精疲力竭，最后在外观的美学上又总是与业主的利润在相互拉扯，有时还得面临大环境的景气问题而牺牲掉细部与建材，最后所呈现出来的作品往往已是各种妥协之后所能呈现相对完整的产物。面对同一块基地，同一个都市纹理，临时建筑所拥有的弹性，刚好能让建筑师有机会更完整地表达看法，也能对建筑有更完整的论述。这种补充说明式的论述延伸，有点像是文本里头的注解，透过注解，去更完整地反应都市与环境，去更完整地彰显设计者的态度与观点。这种注解式的文本，在形式上，反应着建筑设计的时代精神与当代美学，同时反应着一种全球在地化的美学反馈，一种属于城市、市民与设计者的美学想象。在空间的精神上，则是一种延伸，一种对生活空间的品位与态度的延伸，更是一种对未来生活想象的延伸构图。透过接待中心的各种展演与表达，空间的场域存在变成了一种场域连结，像是一种桥梁，连接了现在与未来，实现现在与未来的对话。透过与未来对话的过程，去想象，去勾勒未来生活蓝图。

接待中心的存在，是一种对未来想象的构筑实验。对于地产商而言，整个房产的兴建与销售过程，无疑是一种极度资本密集下的华丽冒险。从土地开发到建筑规划再到市场销售，对于业主，建筑是一种商品，对于设计者而言，建筑却是一种生活态度，是一种场所精神。如何将空间赋予一种生活的品位与态度，透过样板的实验与呈现，作为设计者一种完整的空间论述与美学表达，同时作为地产商后续营造上的检讨与修正调整，在此基础之上，设计者是非常有机会将空间价值转换成价格的一种连结机会，透过空间的实验过程，也是一种不断教育业主的过程，透过这个过程，业主会发现，好的空间坚持，是可以反应在实际的销售数字上的。

接待中心的环境思考，反映出设计者的一种社会责任。接待中心与样板房作为房地产销售的主要道具已成为两岸特殊的商业模式，在全球化的推波助澜下近年来更影响了新加坡、柬埔寨、越南等地区。在早期，重金打造一个金碧辉煌的空间环境，不断用名贵建材与家具堆砌而成的售楼中心比比皆是。在台湾，接待中心的寿命通常也只有阶段性的一年半载，大部分的项目通常在任务结束之后都无回收的可能，几乎就像垃圾一般的被丢弃，如此巨大的废弃物从某个程度上来说对于环境绝对有一定的冲击。在环保意识逐渐抬头的今天，就实际的执行层面来看，环保建材的使用对于成本上的增加绝对是执行过程上的一大挑战，设计之初的美意最后往往抵不过成本上的考验，最后通常只

能以妥协收场。所以，如何在这天平上找出一个平衡点或突破点，是一个设计者对于环境与社会责任绝对该有的一种态度与自我期许。尽管着力点有限，但还是有许多机会点可以透过设计方法来改善。例如，从基地的配置来说，考量基地的永续资源与环境友善，对于原有自然环境的尊重就是一个很好的切入点。例如原有的植被，特别是大型乔木等如何保留或移植，原有的地形地貌如何透过施工法的调整来尽量减少对原有地貌的破坏等。其次，常在接待中心里出现的空间项目，如朝向与日照探讨、屋顶的集水与排水系统、水池与绿化等景观工程、照明与空调的能耗探讨等属于建筑物里环境的考量，不仅能实质性地减少碳排放量，同时能为业主省下可观的成本费用。再次，在空间材料的使用上，如何利用设计的手法呈现项目的品位与独特性，而不是过多的以高耗能的建材堆栈。或者，接待中心在规划设计之初的考量，是否可能在将来作为社区的一部分使用，例如会馆或公共设施等空间，如此一来不仅可以将预算活用，更可以避免过多的资源浪费。另外，在销售过程中的空间分享，例如留设公共性的内部空间供附近居民使用等，诸如此类的环境友善考量，都是一种很好的潜移默化的示范，不仅是一种对消费者无形的教育过程，更能凸显地产商在整个商品化过程中的社会责任，提升企业形象。

"逻辑是无庸置疑，不可动摇的，不过它经不起一个人求生意念的考验。"（弗朗茨·卡夫卡《审判》）。正如卡夫卡在书里所提到，反应在设计的领域中，亦如前文所提及的，整个执业环境对于建筑设计者而言是辛苦的，是不容易的。然而，走了几十个年头，整个大环境的景气问题与房地产业的潜在逻辑规则，以及接待中心作为销售的主要道具，乃至于对于设计者最自由的舞台，从百家争鸣走到当今瓶颈的十字路口，的确也已经到了该重新定义的时候。本质上而言，接待中心的领域本身就是一个竞技场，过程就是一种进化论。也许，生存意志的能量往往能在最严峻的环境里绽放花火，也许大环境的考验，更能淬炼出跨越时代的设计。也期待不久的将来，在接待中心与样板房的领域里，能看见更多进化的作品，更多反应社会文化、当代美学与永续环境的设计作品，甚至，为这个领域带领出一个新的视野与格局。

PREFACE
序言2

售楼处作为一种"复合式社会性空间"模型的概念与实践

齐物设计事业有限公司
总监 / 甘泰来

2015年第十届中国国际建筑装饰及设博会2014—2015年度十大最具影响力设计师（会所空间类）
得奖作品：天晋II会所
台湾室内设计专技协会 2014台湾陈设艺术菁英
2010年第八届现代装饰国际传媒奖 年度商业空间大奖
2010年The 8th Modern Decoration International Media Prize, Annual Commercial Space Award
得奖作品：Level 6ix、BELLAVITA

以一般空间的分类而言，售楼处一直是相对特殊的空间形态，既是"公共空间"，亦是和居住环境议题相关的"展示空间"，而其以销售空间（作为一种产品）为主要目的之企图，更带出其"商业空间"属性的本质。所以，售楼处确实是多种空间类型属性的一种综合体，因此也一直是开发商、销售团队和设计师，对开发项目进行表达，展现开发商团队品牌、项目理念愿景、空间产品形态特色和空间创意等面向上极为重要的发声舞台，更是一个在实体环境中和客户端或社会公众直接互动上最重要的平台。

但随着房地产市场的变动发展，售楼处空间的呈现似乎普遍倾向停留在风格美学或形式造型上的堆砌和拼凑。近年来，我们的观察和体会是售楼处就其空间属性本身的发展而言，或是因应市场持续不断变化的状态，售楼处空间除了可以在风格造型面向的发展精进外，在

图01

图02

空间策略及其空间计划内容(program)上，其实还有很大的潜力面向仍未被积极"开发"。换句话说，就其"商业空间"的本质，在风格或造型上的包装和雕琢，本是无可厚非，也是此种属性之空间在呈现上的重要基本面向，只是进一步地分析，综观售楼处的诸多空间类型本质(商业+公共+展示)的同时，实可让售楼处在当代文化脉络下和城乡实体环境中，作为一种"复合式社会性空间"(multi-functional & social space)模型的概念与实践，于是，有关于空间策略和空间内容上的创新或创意组合，将可能是未来更具新意的趋势，或可具体且多元发展的关键。

所谓的"复合式社会性空间"，可分为两个面向来观察，一个是复合式空间(multi-functional space)，另一个则是社会性空间(social space)。

复合式空间意指空间计划内容(program)上的复合。也就是说，在原本售楼处基本功能空间内容的基础上，因应个案的属性特色，或开发团队的企业形象，甚或该团队(或项目)品牌塑造之企图，复加结合一个以上的(但重点不在多，而在于巧且能够功能化)，原本不属于一般售楼处的功能内容，或将某一原本仅是基本功能的空间内容转化升华，并在空间功能规划布局策略上，赋予其在空间整体组织架构中新的任务位置，让此复加融合或转化的新空间内容，在既不破坏亦能满足原本售楼处之基本功能需求的同时，扩张延伸售楼处在空间活动内容上的影响力，而非仅止于造型风格对视觉感官上的短暂吸引，还能发挥更多元的相关活动"体验"上的互动可能性，进而带出售楼处空间新的能量和生命力。

由于在售楼处中建立了复合式多功能空间，便可在此定期或不定期举办各式配合项目直接相关或非直接相关的活动(例如：艺术展、主题讲座、茶酒会，甚至轻食厨艺和西点烘焙教室或将场地租予精品品牌举办发布会等)，即便仅是售楼处空间的一小部分，也足以成为人们交流和体验的文化活动场域，让售楼处因此升华成为一种能与特定目标客户族群或更广泛族群的互动空间，并且具备更多元可能的"社会性空间"。

以我们执行完成的三个项目案例来说明，一是2014年在台北的"达丽信义"售楼处，由于其坐落于都市环境相对拥挤的巷弄中，且

图04

主要诉求客群是附近区域的换屋客,所以用"邻里间的树屋"为概念,将主要售楼功能空间置于漂浮在二楼高度的木盒体空间中,而一系列木盒树屋下方(地面层)相对开放流通的空间,一方面形成和周围环境间宽敞舒适的尺度关系,另一方面于一侧布局成门厅,一侧则是微型咖啡店。在空间策略上,地面层的门厅和咖啡店可彼此共享,亦可各自独立运作,达成弹性多功能的复合式空间(图01)。或可不定期举办小型茶会、酒会或展览,或就像是邻里社区中,街角的一间可轻松休憩或看书,或聊天的咖啡店。换言之,它既满足售楼处本身的需求,更可独立运作服务周围邻里或随机过路客的微型咖啡店之安排,则是一种空间计划内容(program)上的创意置入,而且是因本项目属性特色而客制化的空间策略和空间内容演化,并非"为咖啡店而咖啡店",是让身为售楼处一部分的"咖啡店",作为可无形延伸并扩大互动效应的触角,形成可与邻里更亲近交流互动的"社会性空间"。

另外两个皆是完成落实于上海的案例,分别是2015年的"明园涵翠苑"售楼处和"信义嘉庭"售楼处。两者皆有持续关于"复合式社会性空间"的发展和尝试。在明园一案,我们置入融合的空间计划内容是一含有类装置艺术式的影像森林(图02)的客制化多功能美术馆(图03),一方面以此呼应项目小区之室内外空间,皆有设置许多当代艺术家的雕塑作品或装置艺术,另一方面则彰显开发商长期在当代艺术的积极推广和支持的企业形象,更可供其展示各式多样的当代艺术收藏。在信义一案中,则规划布局了一处称为"社区营造展示中心"的多功能场域(图04),借由此空间计划内容的建立,让售楼处突破仅是图文影音单向式的讯息展示,而扩及至未来小区活动的示范展演和客户参与体验的多元呈现方式。有关这两个案例的详细内容,读者可参阅本书中的刊载,在此就不再赘述。

在这三个案例中,再次特别强调其中一项重要的共通性,即此复合式社会性空间的概念策略,是基于客制化的角度来切入考量的,并非为了艺术而刻意艺术,而是一种除了风格造型外,在空间计划内容上的衍生和进化,借以升华售楼处在营销推广上的多元性及灵活度,亦开启售楼处的商业面向与人文精神融合的新局面。事实上,市场本身从来就没有对或错的问题,它是一持续不断混沌变动的客观脉络和现实存在,而空间的商业化也从来就不会是设计创作上的阻碍,一切都看创作者如何以开放的心态,不断进化的概念、策略和技术去切入及融入这个永远在变化的市场。我们早已进入新一波的全球化时代,过去的地域环境和人文社会背景虽依然深植于我们的文化脉络中,但面对多元动态复杂的未来新世界,我们不应停在某处,而是尝试更多面向的互动以及发展不同的概念连结,进一步寻觅出一些更具人性和世界性的方向。

齐物设计事业有限公司
Archinexus

成立于2004年;公司中文名为齐物,意取《庄子·齐物论》中万物等量齐观的精神;而英文archinexus,则泛指所有与建筑和设计相关的事业。总监甘泰来致力于发展创意空间设计,2002年自纽约返国,甫在台北初试啼声,其创作的空间设计即获得颇高评价与诸多的回响,由于其在建筑学科的背景及优异的创意设计能力,使得齐物设计在空间设计上始终带有不同凡响的空间魅力与新意。由于公司总监甘泰来开阔而多元的性格,公司除了以空间设计为核心业务,并积极发展相关设计领域之服务项目及产品,本着开放之精神朝多元化经营,以期在创意空间设计产业开创更多的可能性。

图03

PREFACE
序言3

也谈设计的回归

撰文：上海曼图室内设计 / 张成斌

近来，大家纷纷都谈起了设计的回归。

回归是一个有方向性的动词，所谓"回"，总要向着一个方向，而这样的方向又最好是曾经存在过的某种价值体系。

可这样的体系存在吗？即使曾经存在过，又如何保证我们回得去呢？答案似乎又渺茫起来。

我们暂且将这件事情放在一边，尝试着再聊聊别的看是否有所助益。

作为一个设计行业的从业者，我们是很"有幸"地经历了过去二十年中国的巨大变革，可以说这种变革在人类历史上是罕见的。每每经历一次，都是刮目之感，又每每时过数载，总有相认之疑。

有一些事是可查的，2000年时，我们的人均GDP约为1 000美元左右，2014年这个数字最少也会超过6 000美元。并且已有8个地区超过了10 000美元。而又有一些事，却是无法量化观察的，如石阶路变成了8车道，农田变成了小区，早餐铺成了东方既白。

这让中国社会的每一个人都多少有些不知所措，当然也乐在其中，其中包括我们这些从事设计工作的人。

迅速甚至于接近疯狂的巨变，让我们来不及审视，来不及等待，甚至来不及有哪怕是片刻的快乐或忧伤。

大概是因为太快的缘故，当这种发展的脚步稍微停滞，便会让人有种如在三伏天突然停电而失去空调般的不自在，而这样的停滞，大概又是终究要发生的。

也许当这种停滞的插曲奏响时，我们便听到了隐约从各种旮旯里飘来的"回归"之音。

我们总是和建筑及相关产品打着交道。在过去的二十年里，这些产品中的某些特性在被剧烈地强化着。哪一个建筑或者设计具备着哪些特性呢？我想这是一个设计原理的问题，这里也不指望能和所有人达成共识，因为有些问题，但凡不能用量化精确描述时，便不能纳入"科学"的范畴。而非科学的事情，便很难找到什么标准答案。

戏剧性的是，设计也许是世上为数不多的同时拥有"科学"和"非科学"两种属性的。某些时候可以说它"对"或"不对"，某些时候又可以说它"好"或者"不好"。结构不对，建筑开裂，那一定是"不对"，空调不足，新风不够，那也是"不对"，当它满足了结构的稳定，又过于"超静定"而导致了造价的过高，这似乎又是一种"不对"。我们发现，这种"对"与"不对"的标准，虽然是由相关部门的相关法规或规范来衡量的，却也随着时间在不停地变化，有时候，这种变化往往又是滞后的。这时候，"甲方"出现了。我们好多年的工作，就是在"政府""甲方""乙方"这三个团体间发生的故事。甲方是另一把标尺，这把尺上有双重刻度，同时标示着"对不对"和"好不好"的标准。当然，"好"与"不好"没有"对"与"不对"这样容易判断。总之，有种学问叫"平衡"。

那如前面所说设计中的哪些特性在被剧烈性地强化着呢？很不幸，恰恰可能是这种"平衡"。这种"平衡"极有可能是一种无奈的"平衡"，它不同于古人说的"止于至善"，也和"中庸"扯不上半点关系，因为平衡就是平衡，是效率的要求，是资本的要求，是甲乙方关于设计费和现金流的要求，是政府对容积率及地价的要求，是各种要求的无奈平衡。笑用一句佛经用语，"云何平衡，即非平衡，是谓平衡也"。

在这种无奈之下，我们开始了二十年的加速度。是加速而非匀速。

现在看来，"回归"二字依然找不到任何头绪。

那就让我们尝试着再想下去。我们曾不经意间提到"中庸"。我们这些理科生，是用几个脑袋都别想明白的。但我们也尝试着用一点感情，去勉强感受一下"中庸"两字，发现，原来这时下最讨人厌的两个字，却是带着理想而出现的。具体理想是什么，我们不敢说，但是总有那么一种浓烈的理想若出其中。如同白酒形色如水，即使在瓶中，都会让人联想到那种烈味。退而求其次，不妨将"中庸"两字勉强代为"平衡"二字，便有了这样的一句话：有理想的平衡。似乎已经有了一些影子，尽管还不够，也远不够。

我们发现，全中国的住宅，几乎变成了千篇一律，全中国的室内设计，居然可以用几种类似于学名的语言来概括，从ARTDECO、法式新古典、新中式、后现代、现代等，不一而足，现在想想，不禁让人喷饭。喷完之后，又有点不寒而栗。作为设计工作者们，偶尔有些集体自觉，便是认为自己是在做一个产品，而这些二十年来形同复制的产品，是否准备再复制二十年呢？就住宅而言，我们感受最深的便是人在其中，一举一动都被规划得如此明确，每一件家具都无法跳出几种想象，你"被"做了很多事而不自知。整个时代和社会就像一个巨大的RPG游戏平台，你不得不活在一种事先安排好的情节中，去锻炼武功，打掉各种怪兽，更换各种装备。你必须有一个客厅，客厅中必须有一组沙发，洗澡必须要路过马桶，衣柜必须要在那个角落……当你安排好一切以后，又不得不采购很多东西，便有了从宜家到达芬奇等一系列的装备店。

总之，你不会有任何自由选择的余地，除非你是乔布斯，不在房内放任何一件家具。因为这个故事你只有听到结局的权利，而没有一起玩的权利。

设计者也仅仅是这场游戏的参与者之一而已。多年来，大部分人不是没有理想，而是已失去了理想之力，或者叫"理想无能"。因为理想是要靠一定的素养去激发维护的。回归，便是回归"有理想"。理想和有理想不是一件事。

一个好的设计，首先是带有善意的。我们若要讨论所谓的回归问题，那起码要先回归到善意的设计。

善意的第一个要素，是爱心。

善意的第二个要素，是真实。

善意的第三个要素，是自由和谦卑。

至于爱心、真实、自由和谦卑如何具体体现在我们的工作中，这里未必有篇幅可以详述，但本文仅仅先从一个设计工作者主观角度上浅谈了"回归"二字的必要性和必然性，期待以后可作更多的交流。

CONTENTS
目 录

GRAVITY / SALES CENTER DESIGN XI

A Modern Style 现代风尚

012	台湾国扬翡翠森林接待中心	Jade Forest Reception Center
022	25℃海湾一号营销中心	25℃ Bay 1 Sales Center
028	绿地GIC成都中央广场展示中心	The Show Center of GIC Chengdu Central Plaza
038	绿地长沙湖湘中心售楼处	The Sales Center of Huxiang, Greenland (Changsha)
050	上海信义嘉庭接待中心	Faith and Praise Courtyard Reception Center, Shanghai
066	上海明园·涵翠苑售楼部	Green Park Sales Center of Ming Garden, Shanghai
078	中粮商务公园项目示范区营销中心	The Sales Center of COFCO Business Park
084	台湾全坤威峰接待中心	Peak Reception Center, Taiwan
092	东莞万科中心售楼处	Vanke (Dongguan) Sales Center
098	香港天晋II会所	The Wings II
116	台湾富邦丰泰接待中心	Reception Center of Rich Country
128	长白山中弘池南区项目售楼中心	The Sales Center of Zhonghong Chinan, Changbai Mountain
134	保利佛山三山新城西雅图销售中心	Seattle Sales Center, Poly (Foshan, Guangdong)
138	河南台北晶华接待中心	Taipei Jinghua Reception Cener, Henan
146	郑州绿地中心	Zhengzhou Greenland Center
154	河南平天下接待中心	World-Peace Reception Cener, Henan
160	上海虹桥旭辉办公体验馆	The Experience Center of Xuhui Office, Hongqiao, Shanghai
170	光明华强文化创意产业园销售展示中心	The Show and Sales Center of Huaqiang Creative Industry Park
178	珠江科技数码城销售中心	Pearl River Science and Technology Digital City Sales Center
184	福州信通售楼中心	Xin Tong Sales Center, Fuzhou

188	台湾青田青接待会馆	Green in Green Reception Center
194	厦门宝龙一城售楼部	Bao Long One Mall Sales Center
198	万科未来城接待中心	The Reception Center of Future Town, Vanke
204	台湾温布敦19接待中心	Wimbledon19 Reception Center, Taiwan
210	重庆旭辉乐活城体验中心	The Experience Center of LOHAS Town, Chongqing
220	宁波华侨城欢乐海岸售楼处	The Sales Center of the Happy Coast, OCT, Ningbo
226	台湾三本千晴接待中心	San Ben Qian Qing Reception Center
230	周浦绿地缤纷广场售楼处	Sales Center Riotous Square, Greenland (Zhoupu, Shanghai)
236	乡林山海汇	Collection of Forest, Hill and Sea

B Oriental Legend
东方传奇

242	上海中山润园售楼处	Run Garden Sales Center, Shanghai
248	苏州建发地产中泱天成项目售楼处	The Sales Center of Zhongyang Tiancheng, Suzhou Jianfa Real Estate
254	扬州湖滨名都销售中心	Lakefront Fame Capital Sales Center, Yangzhou
268	保利·阳江银滩N2售楼中心	N2 Sales Center, Poly
274	台湾玺悦会所	Seal-Pleasure Club

C Foreign Country Sentiment
异域风情

282	台湾国王城堡会所	King Castle Chamber
290	上海绿地海珀风华售楼处	The Sales Center of Sea Ample, Greenland (Shanghai)
300	合肥海德公馆售楼中心	The Sales Center of Haide Mansion, Hefei
306	肇庆宝能环球金融中心售楼处	The Sales Center of Global Finance, Zhaoqing
314	深圳金众·云山栖Hill Villas 售楼中心	The Sales Center of Hill Villas

A | Modern Style
现代风尚

台湾国扬翡翠森林接待中心
Jade Forest Reception Center

设计公司：域研近相空间设计　　Design Company: Inheressence Design Studio
设计师：李俊平、康智凯　　　　Designer: Li Junping, Kang Zhikai

本案以老庄思想的"太极"与"阴阳"作为配置概念的出发点，透过汉字的行草概念，挥洒出一实一虚且极具生命韵律的建筑量体，象征着存于建筑与生态环境"虚实"与"共生"之间的绿建筑的可能性。

为使绿建筑成为可能，除了在实体的部分尽量不开窗以减少空调耗能，虚体部分透过节能玻璃满足照明需求，同时，以穿越建筑物的水池作为微地区防洪池来减少间歇性大雨造成的地表径流负荷量。另外，透过另一个廊道连接样品屋，以方便分期开发对建筑产品类型进行调整和修改，进而减少建筑资源的浪费。

售楼部和样品屋作为房产销售必备的场所和道具，起到引导消费模式与空间文化的作用。如何将环保、节能、永续与共生的概念，透过空间的安排直接传递给业主与消费者，让此类"临时性建筑"除了满足短期的销售行为外，同时能兼顾环境友好与可持续的概念，是设计者对于本案的最大期许。

在生态环保方面，减少了空调资源耗能，大量运用节能玻璃、节能照明与可回收材料。在环境友好方面，则特别在建筑中央环抱着一面景观水池，除了景观用途外，同时兼具地区防洪的作用。针对近年

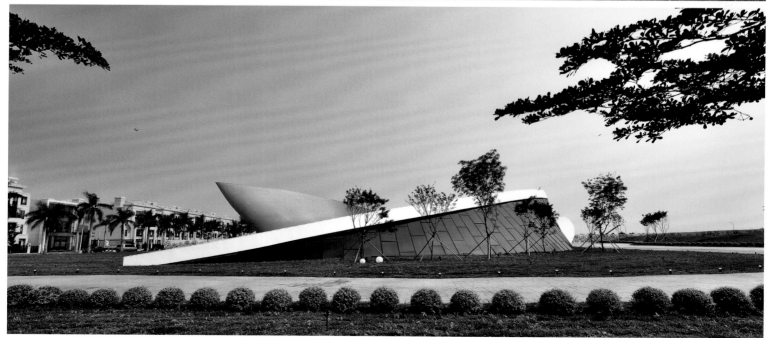

来极端气候的增多，特别是台南地区这种炎热气候雨量不多，但时常出现瞬间暴雨的气候特性，将屋顶的雨水透过特殊的斜度与集水设计，导入水池中再排入下水道，以减少地表径流与局部地区性瞬间淹水的现象，同时兼具分担区域防洪的任务。

特殊的空间形式成功引起地区性的讨论与注意，环境友好与共生的概念与工法亦成功传达了日后建筑物在"绿建筑"工法上广泛应用的理念，对于建筑商的商业形象更是有着显著的提升。更重要的是，透过空间实践的过程，让一向以商业利益为主的房产商，慢慢意识到环境共生建筑的概念，是可以转换成巨大而有形的价值的。

在设计形态方面，本案同样让人眼前一亮，雕塑般的造型，流动的线条，倒映着水面，如同一件艺术品镶嵌在城市的绿地上。VIP区利用降板的方式让客户的视觉延伸到更接近地平线的地方，去感受不同视角带来的不同空间体验。特殊的结构系统设计让整个空间没有任何一根立柱，同时又保持了空间的通透性。减少洽谈区照明耗能的同时，特殊的玻璃节能设计亦能降低空调耗能。

A project this space is accomplished from the perspective of Taoism: Tai Chi and Yin and Yang, where the artistic of concept of running script in Chinese calligraphy has been employed to make real-virtual building volume of vigor and vitality. This symbolizes the possibility of green construction. Coexistence it actually is between building and ecological environment.
Its real part is minimized with openings and cuttings, like window, while its virtual part maximizes the need for lighting with the use of energy-saving glass. The pool penetrates within with an aim to reduce the effect by periodical heavy rain. Another corridor linked to the show flat is of sustainable purpose for feasible modification or alternation to decrease the waste of building resource.
As a necessary tool and field in the sales of real estate, sales center and show flat make guidance for consumption model and spatial culture. The greatest expectation from this project is how to meet short-term sales mission and realize the concept of environmental protection and sustainable development by

现代风尚
MODERN STYLE

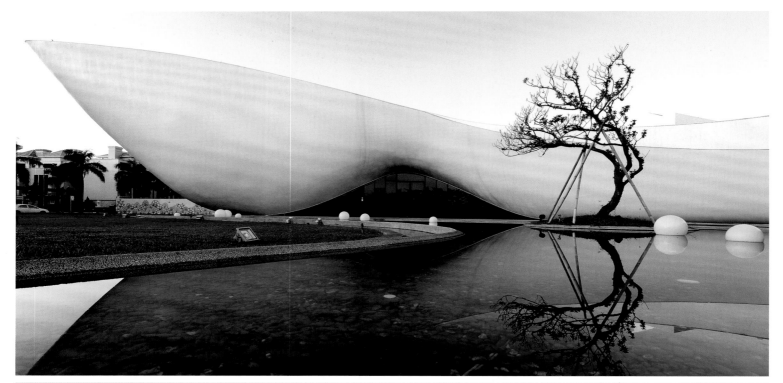

exposing client and consumer to environmental protection, energy saving, sustainability and coexistence.
In this space energy dissipation by air conditioning is cut off to a large extent, and energy-saving glass, lighting of energy conservation and recycled materials are used. The landscape in the middle is both aesthetic as waterscape and functional to control flood, a tangible and measurable response to the local climate, where rainstorm frequents as well as extreme climate. Along the scope, rain on the roof is collected into the pool and then sewer to lessen the ponding.
The special form of this building has drawn regional discussion and attention, while the avocation and the process of being friendly to environment set up a model for wide use of green construction. This is bound to enhance the business image. What's more important is that, such a trial leads to benefit-oriented cooperation to awareness of the importation that the

现代风尚
MODERN STYLE

building and its surroundings should co-exist, by which huge and visible value can be got.

As for its design form, it really makes a refreshing impression: the sculptural modeling, the stream-lined lines, are the reflective water are like pieces of art that have been embedded onto a city greenbelt. The VIP area, treated downgrade, which gets the customer's vision close to horizon to feel different spatial experience from various viewpoints. Thanks to the systematic design, there is no column readily available, which keeps the space open and transparent. In the communication area, energy lighting needs is greatly while the energy conservation design for glass helps to lower the pressure on air conditioner.

25℃海湾一号营销中心
25℃ Bay 1 Sales Center

项目位置：海南三亚
设计公司：李益中空间设计有限公司
硬装设计：李益中、范宜华、黄强
软装设计：熊灿、王雨欣
用材：白色人造石、灰色石材、艺术地毯

Location: Sanya, Hainan
Design Company: Li Yizhong Space Design
Decoration Designer: Li Yizhong, Fang Yihua, Huang Qiang
Upholstering Designer: Xiong Can, Wang Yuxing
Materials: Marble, Artistic Carpet

嘉鹏·25℃海湾一号营销中心采用既充满情景海元素又类似飞碟的模型打造，让业主如同置身时空幻觉之中，又如同置身海底，与深海鱼类和珊瑚群共舞。同时，室内设计紧扣简洁、富于品质感的新现代主义建筑风格，合理分布各功能区域，使之动线流畅，从而使室内空间的最大、最优成为可能。整体空间突破传统设计构思，依据空间布局，通过精简的素材、独特的造型、流畅的线条、简约的配色以及

现代风尚
MODERN STYLE

万有引力·售楼部设计
GRAVITY / SALES CENTER DESIGN

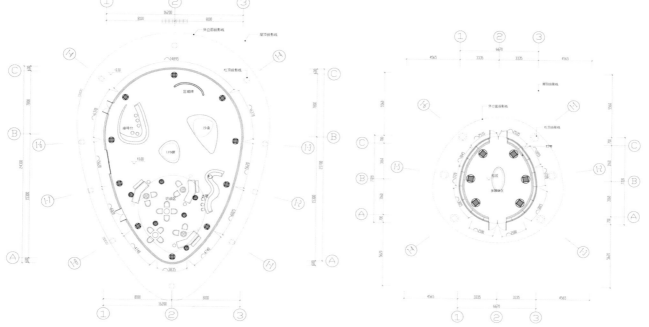

微妙的灯光处理，打造出一个极富创意和个性的超酷场景，借此诠释出现代与活力的城市精神。

这个宽敞的空间，以白色为主导，局部点缀以绿色（草绿），无处不体现着现代室内空间的开放性和流动性。设计对于媒材的运用也极为考究，以清新、亮丽、纯粹为追求，尤其是对玻璃的使用。玻璃铺设于整个营销中心的墙面，通透的属性有利于将室外绿意美景及日光纳入室内，打造出与绿景共缠绵的山居生活。

造型的设计，是此次营销中心设计的重中之重。在天花板的装饰上，设计师独具匠心，利用"飞碟"圆润流畅的线条，凸显整个界面的凹凸有致，并采用巧妙的灯光效果渲染，使天花板像浮起来了一样，显得更加轻盈优雅，给予来访者科技感爆棚的体验。接待台、沙盘、LED屏、吧台以及家具也同样是不规则的造型，给人炫酷的、独树一帜的感觉，使空间充满了生机与活力。而这一切，其实都归功于设计师丰富的想象力和创新理念。

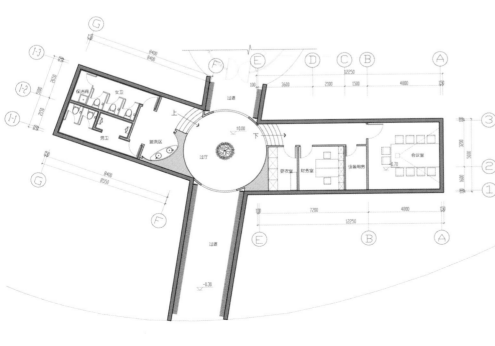

This is a project done with sea elements for visitors to feel as if in the depth of sea, dancing with fish throughout coral. Its interior is closely related to the modernism, succinct but of great quality. All functional areas undertake reasonable distribution, so traffic lines go smoothly to maximize the spatial potential. The holistic breaks away design stereotype. On the basis of the space structure, materials are being used simple, modeling is unique, lines are fluent, hues are contracted and lighting are marvelous. A scene comes innovative and individualized to make annotation of the city spirit: modern and vigor. The large and wide space is dominated with white and grass-green interspersed in some parts. Everywhere is the openness and fluidity that's exclusive to modern interior design. Materials are carefully chosen, with significant emphasis on freshness, brightness and purity, particularly the glass which is used all over walls to take green within to make a life lived in hills and waters.

The salient feather is the modeling given priority and priorities. The ceiling maps out the lines of flying saucer to protrude the concave-convex interface. With light effect sketching, the whole ceiling seems to be floating. Lither and more graceful it is, from which viewers feel a strong sense of high tech. Likewise, the reception desk, the sand table, the LED, the bar and the furnishing is of irregular appearance, cool and special to inject into the space vigor and vitality. And as it is, such a package is done with the imagination and innovative ideas of the designer.

绿地GIC成都中央广场展示中心
The Show Center of GIC Chengdu Central Plaza

设计公司：赵牧桓室内设计研究室
设计师：赵牧桓
参与设计：安娜、戈梅斯、雷玉荣、赵自强、李欣蓓
摄影师：黎威宏
用材：玻璃纤维、加强石膏板、木材、PVC地板、不锈钢、大理石、白色涂料
面积：2 700 ㎡

Design Company: MoHen Chao Design Assoc.
Designer: Zhao Muhuan
Participant: Ana Patricia, Castaingts Gomes, Lei Yurong, Zhao Ziqiang, Li Xinbei
Photographer: Li Weihong
Materials: Glass Fiber, Reinforced Gypsum Board, Wood, PVC Flooring, Stainless Steel, Marble, White Painting
Area: 2,700 ㎡

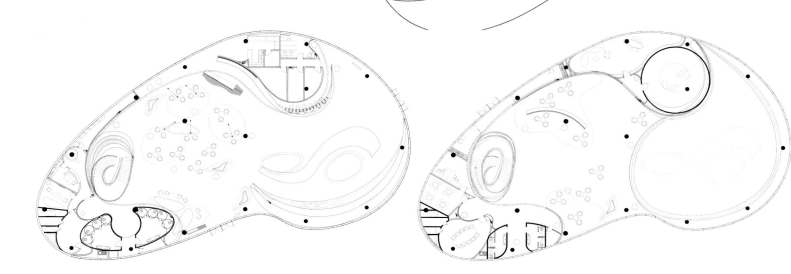

绿地 GIC 成都中央广场展示中心是一个展示空间，位于成都市的中心地带。

该项目的设计突破了传统设计的表现形式。设计初衷是想让访客通过室内细部与建筑外部由内而外的互动，利用自然和艺术的表现手法，营造出一种现代和可持续发展的城市印象。该建筑有其感性的一面，再加上与之协调的景观，使之成为一座艺术感和建筑表现兼具的现代建筑。律动的几何形状由建筑外立面延伸到室内空间，统一的造型贯穿着景观、建筑和室内设计，整体空间具有不间断的飘带状线条感。螺旋式向上旋转的楼梯造型将整个空间分隔成不同功能的区域。

设计师以水和自然为设计灵感，打造出一个流畅的、不间断的动态环境空间。

空间本身具有功能性和雕塑性，在上下两层整体共 2 700 平方米的空间里营造出不同的区域，如大型展示厅、接待台、酒台、休息室、商务贵宾区，在这里你将沉浸于极具艺术感与雕塑感的家具陈设之中。

现代风尚
MODERN STYLE

Greenland GIC Chengdu Central Plaza is an exhibition space, and located in the heart of Chengdu.

The project is aimed to exhibit real estate properties in a no traditional way, with an intention to give the visitors an experience of interaction between the exterior and interior with natural and artistic approaches to create a impression modern and sustainable urban development. The building with its sensuality and harmony with the landscape makes a symbol of an artistic and architectural expression. The geometry of the exterior facade is extended into the internal, connecting landscape, architecture and interior design. The holistic space is positioned as a continuous ribbon, where the screw-up stairs go upward in dividing all into different functional sections.

Water and nature work as the main inspiration of this project, achieving a fluid space in a smooth uninterrupted and dynamic way.

With inherent functionality and sculptural sense, the space has a sequence of different areas in 2,700 sqare meters onto its two floors, like the large showroom, the reception, the bar, the lounge, and the VIP areas, where you feel nothing but immersed in pieces of art and sculptural furniture.

万海有引力 售楼部设计
GRAVITY SALES CENTER DESIGN

万有引力·售楼部设计
GRAVITY / SALES CENTER DESIGN

绿地长沙湖湘中心售楼处
The Sales Center of Huxiang, Greenland (Changsha)

Design Company: G&A Design
Participant: Wang Yiwen
Photography: Three Images / Zhang Jing

本案的设计理念是想要创造一个自由而且流动的空间，在这些空间中充满具文空活动区域，上海及湿原叠垂常人性化的绿植，让绿色散设计师每英尺的身泛绕建筑的平面有着不规则的形状，这种无定型的中央核心必要比传统的排列复杂得多。

州万物起源的混沌到未来的模糊不着，这两极之间由无数个片段与瞬间组成。人们的情感里总会有一份对于体验记忆的眷恋，那是人们觉得属于自己情感的美好部分。

人们根据活动私密程度赋予空间不同的属性，使之联动并相互渗透，使得边界模糊。"山墙"的间隙大小使空间因不同情态而产生的变化，有隐秘的，有豁然开朗的。许多线性垂直相交的装置构筑成"叠加的墙体"，在灯光的作用下透着斑斓的光影，在这些界面中如自然的树影婆娑。粗糙的材料与光滑的金属贯穿在不同元素间，使原始与未定型的中间的因素在同一个空间发生趣味的碰撞。

从万物起源的混沌到未来的模糊不定，这两极之间由无数个片段与瞬间组成。人们的情感里总会有一份对于体验记忆的眷恋，那是人们觉得属于自己情感的美好部分。

本案的设计理念是想要创造一个自由而且流动的空间，在这些空间中充满了社交活动区域，以及装点着非常人性化的绿植，让绿色散布在每个人的身旁。建筑的平面有着不规则的形状，这种无定型的中央核心要比传统的排列复杂得多。

人们根据活动私密程度赋予空间不同的属性，使之联动并相互渗透，使得边界模糊。"山墙"的间隙大小使空间因不同情态而产生的变化。有隐秘的，有豁然开朗的。许多线性垂直相交的装置构筑成"叠加的墙体"，在灯光的作用下透着斑斓的光影，在这些界面中如自然的树影婆娑！粗糙的材料与光滑的金属贯穿在不同元素间，使原始与未来的因素在同一个空间发生趣味的碰撞。

万有引力 售楼模型设计
GRAVITY SALES CENTER DESIGN

What's from the origin of... bounteous with green... where you are. The irregular of future consists of num... form of the space... more difficulty than that by between such two poles. Inside the... those traditional... are sentimentally attached to some experience... sections are... with differentiated properties based best part in human memory actually. on private a... or them to be interlinked and penetrated This project is aimed to create free and flowing space, with their boundary consequently blurred. The gable gaps

万有引力 售楼部设计
GRAVITY SALES CENTER DESIGN

offer the space more contextual changes. Some are in dark while some are open and clear. The vertical-crossed devices altogether make superposed walls, which bring forward multicolored shadows with light cast on. The shadows seem to have been accomplished by trees dancing in the breeze. Throughout various elements are coarse materials and glazed metals. Between elements of the primitive and the future is an interesting collision that is happening under the same roof.

上海信义嘉庭接待中心
Faith and Praise Courtyard Reception Center, Shanghai

设计公司：齐物设计
设计总监：甘泰来
参与设计：魏振铭、简伶捷、吴旻熹、张瑞成
摄影师：卢震宇
用材：石材、木丝水泥板、铝板喷涂、木皮、镜面不锈钢、墨镜
面积：1 850 m²

Design Company: Archinexus
Director Designer: Gan Tailai
Participant: Wei Zhenming, Jian Lingjie, Wu Minxi, Zhang Ruicheng
Photographer: Lu Zhenyu
Materials: Marble, Cement Board, Sprayed Aluminium Sheet, Veneer, Mirror Stainless Steel, Dark Mirror
Area: 1,850 m²

本案空间以"山、水、云、瀑"为形式概念，于各区域中转化成形，借以呼应项目建筑外观立面之设计主题。在平面功能布局上，设计以一楼大堂为中心，一侧为"社区营造展示中心"，另一侧为"洽谈区"，再搭配二楼的 VIP 签约区。

"社区营造展示中心"，是一个既可并行融合于售楼处其他基本功能空间的场所，同时又是一个可以独立运作的多功能空间场域。在此场域，一方面在常态上有着项目社区内容的展示，另一方面亦是举办相关实质主题活动的场所。换言之，所谓"社区营造展示中心"的概念和实践，乃是企图除了以图文的展示方式外，让以后的社区活动能有更实际展现和体验，让售楼处的项目展示内容，跨越影音图文式的"讯息展示"方式，将以后小区关于人文、亲子、邻里等的"活动展演和实质体验"置入售楼空间中。透过这种多功能式空间的融入，让小区"软件式"内容得以呈现，甚至可以是不定期式（和项目并非直接关联）的主题活动或展览，使售楼处不再局限于单向的讯息传达，进而升华成一个具备活动体验及互动的社会性空间。

在"洽谈区"部分，以图书馆塑造其空间意象，除了传达小区的人文气质外，更将展示物和洽谈区座位彼此交错布置，互相交融，突破传统上洽谈和展示分明的分区方式。于是，项目模型成为图书馆中的收藏，图书亦是图书馆的藏书。在此项目中，销售不再是唯一目的，社区人文的营造和推广，才是更大的企图和目标。

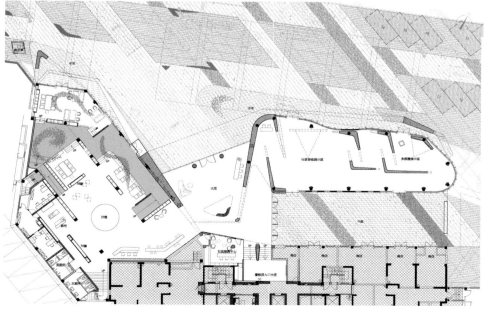
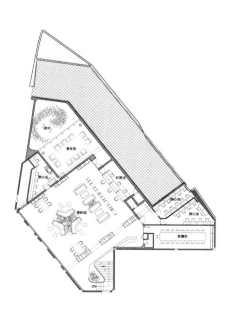

现代风尚
MODERN STYLE

The concept of "hill, water, cloud and waterfall" has been concretized in sections to correspond with the theme of the building facade. With the 1st-floor lobby as the center, the center of community display and the conference areas are flanked on both sides. The room for VIP is fixed onto the 2nd floor.

The center of community display serves both for basic functions like sales promotion and multi functions, where activities relevant to community and other themes can be hold. In other words, apart from the image-text display that can be finished besides the usage of video, image and text, activities that occur in the community can take place ahead of time, like those on culture, the relationships between parents and children, and among neighbors. The fusion of the multi-functional space realize what's on in the community in form of software, or rather theme activities or exhibition can be witnessed now and then. The sales center is not a carrier to convey message in a unidirectional way, but upgraded into a social space for people to participate, to experience and to interact.

In the communication area, library is used to shape the spatial image to express the cultural temperament exclusive to the community and to stagger the exhibition and the seats in the communication. So the clearly-cut division of communication and display is broken away. Sand table becomes the collection in the library. So do the book. Sales has no longer been the sole purpose, and more important is the creation and promotion of the community culture, a greater attempt and purpose.

上海明园·涵翠苑售楼部
Green Park Sales Center of Ming Garden, Shanghai

设计公司：齐物设计事业有限公司
设计总监：甘泰来
参与设计：魏振铭、黄怀萱
摄影师：卢震宇
用材：石材、木皮、茶镜
面积：1 433 m²

Design Company: Archinexus
Director Designer: Gan Tailai
Participant: Wei Zhenming, Huang Huaixuan
Photographer: Lu Zhenyu
Materials: Marble, Veneer, Tawny Mirror
Area: 1,433 m²

现代风尚
MODERN STYLE

本项目的设计概念为"层山郁林",借以呼应业主开发案的"森林都市"主题。

层层的山势被转化为空间与空间、区域与区域之间,一系列斜切的块体或墙体,看似彼此相互遮掩,实则是在错落有致的关系中,塑造或压缩或展开的空间,交替变化的层次转换,亦作为视线场景的框景,及行进动线的引导暗示。

在一楼近中心处,大胆置入的斜切量体,自然形成出入口大堂之当代艺术展示区和斜切量体之后的项目影像展示区。而斜切量体的内部,刚好成为一楼和二楼间的阶梯动线,以迂回的隧道之姿,将客人引导至二楼的洽谈区。

在斜切量体之后的区域,搭配着空间中原本的结构柱,再加入若干比例不一的假柱,真假柱体的四面以视频包覆,将"郁林"转化为展示项目内容的"影像森林",突破传统单一影音室的展演方式。更进一步地,"影像森林"的播放内容,亦可配合一楼举办的与艺术相关的主题活动来进行切换,于是,一种融合于售楼处中的多功能式空间形态浮现成形。

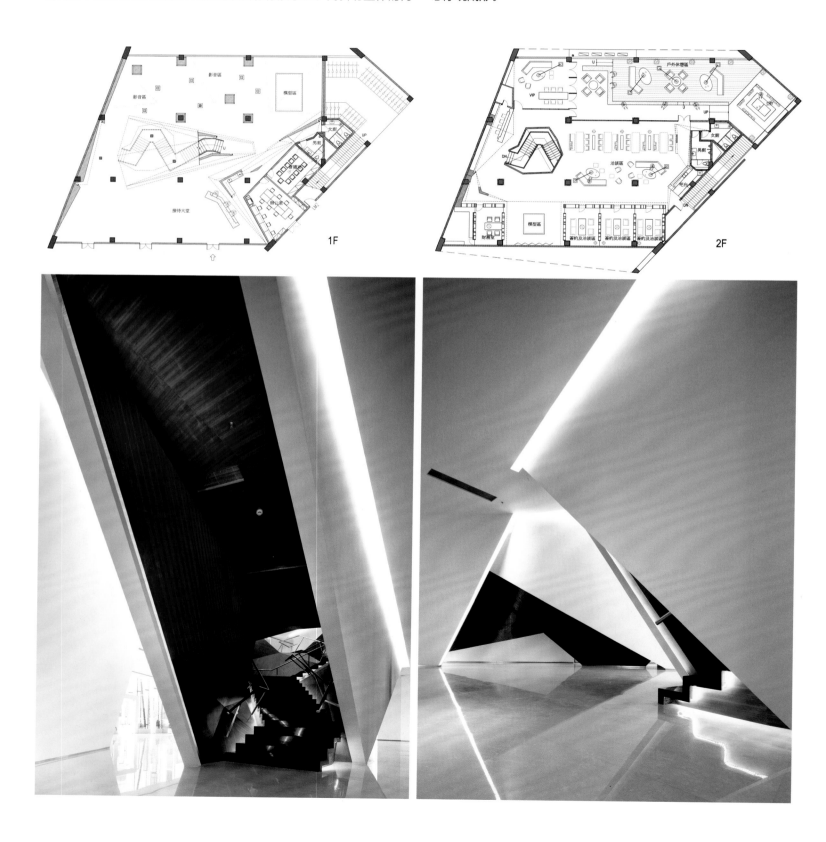

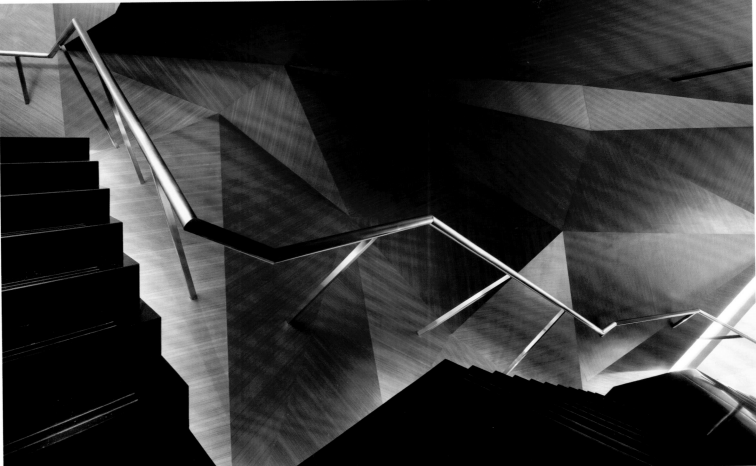

万有引力·售楼部设计
GRAVITY / SALES CENTER DESIGN

A project this space that takes hills and forest as its design philosophy to make good correspondence with the theme of urban forest initiated by the developer.

Hills have been shifted into space: between sections a series of slanting block and walls, reciprocally hidden on the surface but actually compressed or unfolded in the well-proportioned relation while serving as the frame picture to guide traffic lines.

Around the middle of the 1st floor, a beveling volume is bolded fixed to give birth the displaying area of art and the image show sector behind the chamfer. As for the volume inside, the stairs goes upward into the 2-floor communication area.

Behind the beveling modeling are rows of structure columns made of log, which with virtual ones large or small make an image of forest to break away the traditionally monotonous form to display. Additionally, contents of the "image forest" can be used feasibly and flexibly according to the artistic theme that's being held downstairs. A multi-functional displaying space has consequently been in a sales center.

万有引力・售楼部设计
GRAVITY / SALES CENTER DESIGN

中粮商务公园项目示范区营销中心
The Sales Center of COFCO Business Park

项目位置：广东深圳
设计公司：李益中空间设计有限公司
设计师：李益中、范宜华、陈松
软装设计：熊灿、刘灿灿、欧雪婷
用材：白色人造石、爵士白大理石、檀木黑大理石

Location: Shenzhen, Guangdong
Design Company: Li Yizhong Space Design
Designer: Li Yizhong, Fang Yihua, Chen Song
Upholstering Designer: Xiong Can, Liu Cancan, Ou Xueting
Materials: White Artificial Marble, Jazz White Marble, Black Sandalwood Marble

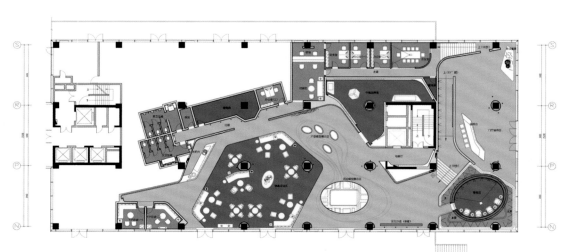

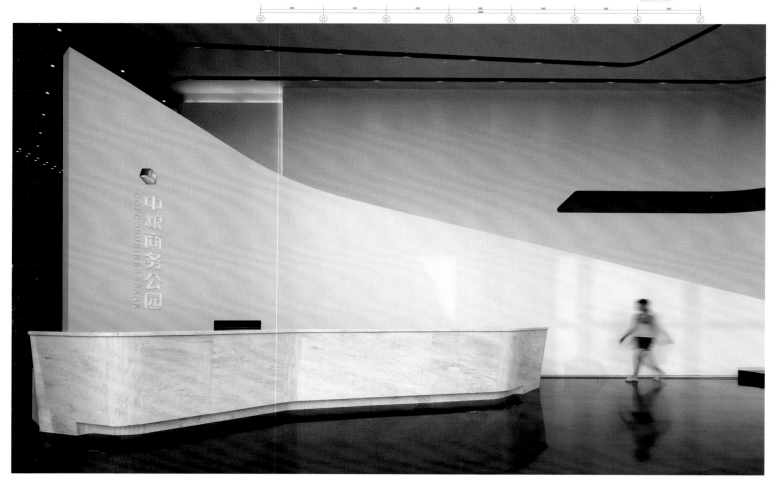

中粮商务公园位于宝安创业二路与留仙二路交汇处，地处深圳前海湾一级辐射区，属于政府重点打造的北硅谷核心区域。它拥有迅捷的立体交通，15分钟直抵深圳宝安机场，20分钟地铁直达深圳北站。项目占地3.6万平方米，建筑面积10.8万平方米，集产业办公、商业、居住为一体，是集新型产品研发、展示、企业孵化、专业人才培训、商务服务于一体的多功能产业综合体，重点发展包括科技研发、创意产业、服务外包等在内的"双高型"产业。

这样一个创智未来空间，其气质和气度，要看它赋予空间的想象力和生命力，空间里弥漫的是友好还是冷漠，是安定还是恐慌，是保守还是激进，它可能包含的是各种基因，都在它的空间里袒露无遗。而设计师，作为空间陈设的装扮者，同时也是空间灵魂的感受者。他们寻找不一样的态度，面向未来，创造智富。于是，于本案的设计中注入了形与色，与空间交相辉映，迸发着激情和能量。

白色与黑色的泼洒，纵横有序的线条的运用，让人充满了无限的怀想。空间中，每一处酷红的运用都化成美妙的音符，洋溢出无限的热情，悦动着每一位来者的心弦。同时，设计更专注于空间的细节和品质，让来访者可以静静地享受这独特的空间魅力。在这里，我们都是自由的天使，享受着不一样的魅力"旅程"。

万有引力·售楼部设计
GRAVITY / SALES CENTER DESIGN

The location of COFCO Business Park at the conjunction of two roads, is a forward position of Shenzhen Front Bay. A core area it is that the municipal government aims to build with efforts. It has a convenient traffic system, 15 minutes to Bao'an Airport, 20 minutes to Shenzhen North Statin by metro. On the site of 36 thousand square meters, it enjoys a building area of 108 thousand square meters while making a multifunctional compound by integrating office, residence, business to conduct research and development, display, enterprise hatching, human resource training and business service. Particularly, "two-high" industry like research and development, creative industry and service outsourcing is given priority.

Such an intelligent space oriented toward future, whose temperament and tolerance depend on the imagination and vigor for it. Whether it's friendly or cold, calm or panic, or conservative or enthusiastic, all possible genes it contains are exposed to the full. And the designers as its dresser or its perceiver, has found a different attitude toward its future to seek after wealth. So the project is instilled with forms and colors that have been complimentary to each other to burst out passion and energy.

The use of white and black, and the lines vertical and horizontal allows for endless vision. Each part in red is beautiful note to ooze nothing but passion and enthusiasm, touching and pleasing every comer. Meanwhile, spatial detail and quality have been paid more attention for visitors to enjoy the special charm as free angels to be indulged in a journey of another kind.

现代风尚
MODERN STYLE

台湾全坤威峰接待中心
Peak Reception Center, Taiwan

设计公司：域研近相空间设计　　Design Company: Inheressence Design Studio
设计师：李俊平、康智凯　　　　Designer: Li Junping, Kang Zhikai

　　本案是一个大型都市更新的推案，地处台北市万华区的老旧社区内，处在短暂的销售期间内的都市街角，从拥挤的老旧街道到豁然开朗的景致，设计师企图在城市营造出柳暗花明的桃源之境，运用流动的量体堆栈，时而由里到外，时而由外到内，穿梭来去，营造出一种阔山后的山林云雾间，如云似雾却又如石似山的层峦绵延的行旅山景。

　　花费了数年才整合成功的都市更新案，需要一种宣示性的新生力量以及强烈的象征语言，来宣告在这老旧市区即将有一个新的生命诞生。以雕塑体的艺术性概念作为街角的城市地景的设计，设计者期待一个临时性建筑物的诞生与消失，被满足的不应该只有销售行为本身，应该同时能兼顾微都市的街道地景任务。即使是昙花一现，都能为这城市的街景，增添些许的惊喜与美感。

　　样品屋作为房地产销售的主要道具已成为台湾独特的商业模式，近年来更成了房产销售必备的消费模式与社会文化。

　　然而这种临时性建筑的存在，除了满足销售考量的商业行为，是否能多一点都市景观上的关注与阶段性的社会文化考量。设计者企图从公共艺术的角度出发，利用建筑外观艺术化的雕塑体来点缀街道空间与视觉景观，同时留设大面积绿地与开放空间，让短期的销售空间也能为这忙碌的城市增添一些艺术氛围。

This is a refurbished project in old quarters in Taipei, an urban corner that needs to accomplish a mission of a short-period sales. A project this is that seems to have been rising out of its old setting all of sudden, making an effect that dark willows is against with bright flowers. The accumulation of flowing volumes, some of which are now inside while some outside then, creates an image forest below forest lies behind great hills, where to start a journey: sometime you are going through clouds and sometimes through range upon range of mountains. This is a project that makes its acclaim and validation of new blood and token that has taken several years to integrate these updated success. An artistic concept in form of sculpture has now been towering as a landmark. As is expected by the designer, the appearance of a contemporary and its disappearance should not just have its sales purpose, and instead, it should give more consideration to more things, particularly the landscape of its surroundings. Though transient like a flash in the pan, it can offer surprise and aesthetics for where it is.

The show flat is taken as a main tool for sales in real estate.

A business model is not only popular in Taiwan, but across the world, which has recently exerted a great effect on China mainland in accomplishing a necessary consumption model and indivisible social culture in property marketing.

When the mission of a contemporary building is met, whether it can more or less focus on urban landscape and periodical social culture should be taken into account. From the perspective of public art, the sculptural appearance of this building is made optimized use to embellish street and visual landscape. Meanwhile, quite a proportion of area is reserved for greenbelt and used as open space, so a short dated sales center finally adds some artistic ambiance to where is located.

现代风尚 | MODERN STYLE

东莞万科中心售楼处
Vanke (Dongguan) Sales Center

设计师：李益中
用材：饰面板、喷砂不锈钢板、白色人造石、爵士白瓷砖

Designer: Li Yizhong
Materials: Veneer, Sprayed Stainless Steel Board, White Artificial Stone, Jazz White Porcelain

在设计师李益中以往的作品中，常常理性与感性兼具，追求格调且富于情感的诉求。在人文素养方面崇尚文化韵味，致力于塑造优雅的情境空间，东莞万科中心的售楼处则再次体现出了他的建筑设计理念。

地平线上，远远可见的这栋巨型长方体，如同跃出水面的鲸鱼一般，令人过目难忘。建筑的外立面采用了大量喷砂不锈钢作为主材料，建筑侧面以规则的几何三角形配合线性灯光，营造出抢眼的视觉效果。现代感极强的几何形立面，搭配占建筑总面积近一半的绿色景观，彰显东莞万科售楼处前瞻性的建筑环境理念。

这栋建筑的另一有趣之处在于入口处采用全透明落地玻璃作为玄关，从整体上远观建筑，如同掀起长方体的一端，让人有种迫切想要进入其中，一探空间内部构造的冲动。另一方面，建筑将一楼及二楼的空间通过绿色景观布局，良好地结合起来。步入建筑内部，空间的功能划分一目了然。在入口处便可窥见售楼处的沙盘区、洽谈区，这种敞开式的结构消除了神秘感，取而代之的是一种开放性的大气。

与建筑外立面相同，内部的墙面也采用了几何形的三角模块。不同之处在于内部空间，采用大的体块转折造型来塑造出空间的划分与立体层次，并且在室内顶部以相同的处理手法，营造出错落有致的天花。

在中央部位以巨大的瀑布式水晶吊灯作为点睛之笔，创造出令人惊艳的感官享受。室内空间部分的色彩简洁明朗，天花与地面和白色对应，银色的饰面与原木色相互匹配。

在软装方面除了沙发略带深色，桌椅则同样采用浅色系作为主色调，洽谈区的座椅充满未来科技感，但又不失温馨优雅的气质。

值得一提的是，李益中在建筑内部引入了部分亲水景观，与外部的水景融合起来形成连贯的景观。优秀的建筑是人文与科学的完美结合，更是艺术性的创造，东莞万科售楼中心的内部空间，动线清晰分明且布局恰到好处。户外部分，在整体布局的同时，引入了大面积的绿色景观，从整体面积上看，亲水、绿色景观占到了总面积的三分之一，体现了建筑与自然和谐共生的设计思想。

现代风尚
MODERN STYLE

万有引力·售楼部设计
GRAVITY / SALES CENTER DESIGN

As is the practice in the project by Li Yizhong, it's both reasonable and emotional in the pursuit of style while turning to sentimental appeal and stressing cultural taste in being devout to spatial grace. Such is a project to embody his design philosophy time and time again.

Like a huge cuboid, the building, if looked at from a horizontal perspective, is like a whale jumping out of water. So eye-catching and unforgettable it is. With amounts of sand-sprayed stainless steel as the main material used onto its facade, the sides creates a striking visual effect with regular triangle with linear lighting. So modern is the geometric facade embellished with greenery landscape that almost covers an area as half as the total building area that the sales center presents a prospective constructional environmental philosophy.

Another interesting aspect is the all-transparent landing glass as the vestibule at the entrance. In the distance, the whole building looks like an end of a cuboid that has been raised, which stimulates people's curiosity to get sight into to explore the internal structure. On the other hand, with greenery landscape, the 1st and the 2nd floors have been fused together. Inside, all functions are clearly-cut. Eyesight at entrance can reach the sand table, the communication area, an open structure that has erased the mystery but presented a magnificent openness.

The internal walls remain the same as that of the facade. The differentiated lies in the interior space, where large turning blocks are used to partition the space into solid hierarchies. The ceiling is treated with the same approaches to build up well-proportioned ceiling.

In the middle is the huge waterfall-like chandelier as the finishing point to exert a very stunning effect on sense organs. As for the hue of the internal space, it's concise and bright, with walls and flooring both coated in white, and the silver veneer matching the burlywood.

Except that the sofa is somewhat dark brown, the table and chairs focus on light color. The chairs in the communication area is brimmed with sense of future science and technology, yet equally warm and significant.

What's most remarkable is the introduction of waterscape within, which is joined with that outside. Any good building should make an impeccable combination of culture and science and technology, and more particularly a creation of art. Inside the space, traffic lines are arranged in a duly state. Outside, the large patch of greenery is done on the basis of the whole layout. The one third of waterscape and the green is a concrete embodiment of the coexistence of building and nature.

香港天晋II会所
The Wings II

设计公司：齐物设计
设计总监：甘泰来
参与设计：刘煜铃、张芃欣、刘思吟、武金凤、黄怀萱、魏振铭、林韦伶、张瑞成、郑明安、赖彦铭、范秉钧、陈海伦、何智渊
摄影师：卢震宇
用材：意大利米白洞石、金锈石、欧网石、米黄洞石、黑色镀钛镜面不锈钢、水晶玻璃金属吊件、黑檀木皮钢琴烤漆、杷檀木染灰木皮、柚木木皮、橘色马鞍皮革、纹路壁布、图案定制地毯
面积：2 662 m²（室内）、1 832 m²（半户外）

Design Company: Archinexus
Director Designer: Gan Tailai
Participant: Liu Yuling etc.
Photographer: Lu Zhenyu
Materials: Marble, Black Titanized Mirror Stainless Steel, Crystal Glass Metal Fittings, Stoving Varnish, Veneer, Leather, Wallpaper, Carpet
Area: 2,662 m² (Interior), 1,832 m² (Semi-Open)

现代风尚 | MODERN STYLE

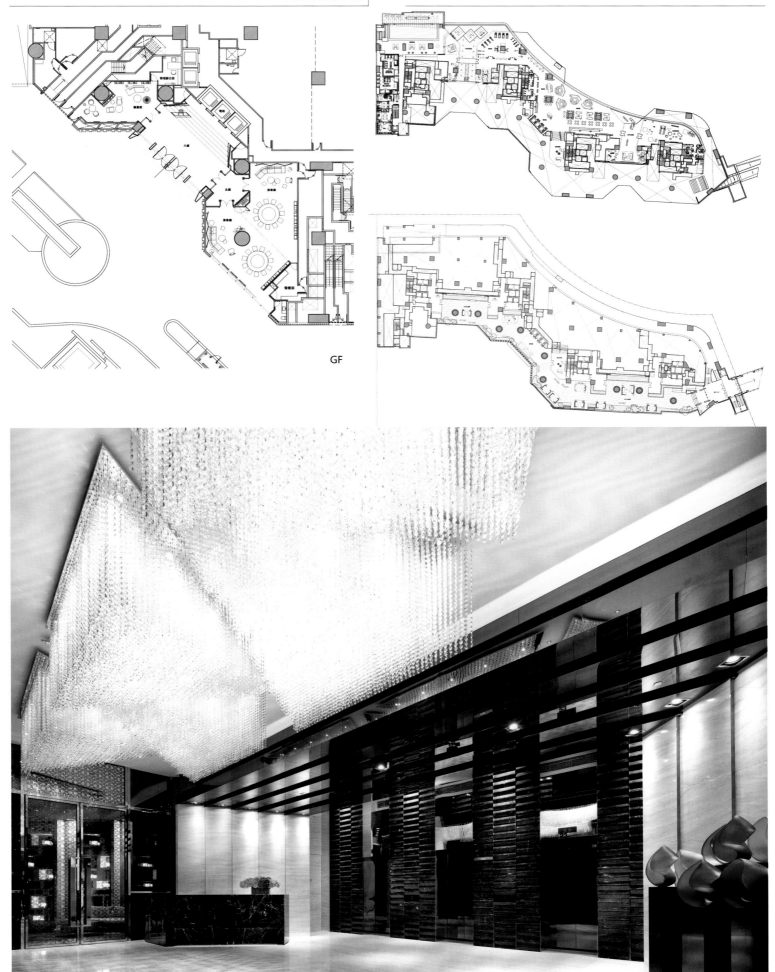

GF

现代风尚
MODERN STYLE

本案设计意在于有限的实体环境中，创造空间感知上的无限性。

在社区住户入口廊道中，一系列错落的大形柱体，被转化成繁茂的大树，透过格栅线条的辗转延伸至天顶，交错编织出充满无限空间层次的都市森林半户外休憩空间。

在会所入口处，借由天、地、墙融合的形式，将梯间转塑成空间框景，望向会所的主题雕塑物件，而此物件是超大型吊灯 (chandelier)，是天顶和行云，亦是流水和瀑布。于是，若干种自然形式与现象的跨界融合，彰显出会所案名——天晋，并在空间感官上及意义精神上无限延伸。

健身房在以线段式 LED 灯，同方向但不规则间距的安排于反射面的吊顶，搭配着大楼玻璃帷幕的加成效果，一种流星雨下的透视空间，引领我们迈向无限的能量境界。

无边际泳池，模糊了水面与陆地的界线，融合了水与陆地的景致。在巡游的过程中，是体能的锻炼，亦是心灵的无限释放。

亲子育乐室是一个被缤纷色彩点缀的摩登树屋。这是孩童的秘密基地，是他们无限潜能开发的新天地。

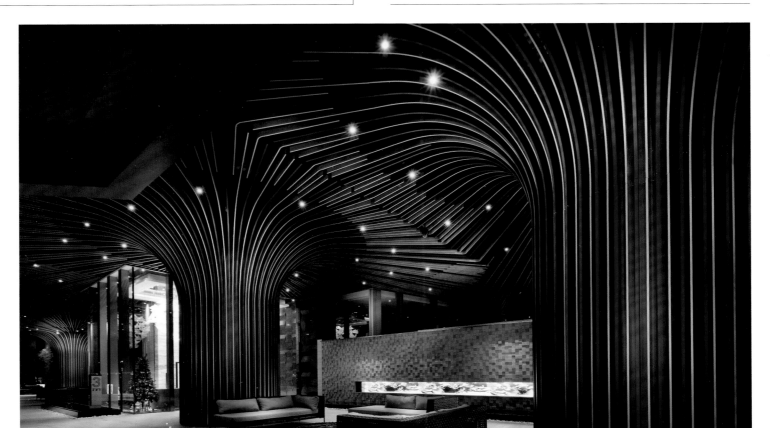

现代风尚
MODERN STYLE

This is a project with limited space unexpectedly generating perceivable infinity.

At the corridor into the community, a series of well-proportioned large columns have been transferred into giant luxuriant trees. When finding their access into the ceiling, the trees are staggered to make a semi-open urban forest as a good resting place.

Around the entrance, the stair well is shifted into a spatial frame with the aid of fusing the ceiling, the flooring, the wall and the physical form. The theme sculptures oriented toward the interior is actually a super large chandelier, implying dome, flying cloud, flowing water and waterfall. Numerous natural forms and phenomena have been fused across fields to highlight the name of the project, Tian Jin that literally means heavy entering in Chinese language to accomplish infinity of spatial sense organ and spiritual level.

The gym is embellished with wire-form LED light fixture arranged in the reflective suspended ceiling in the same direction but at irregular space distance, which with the glass curtain to make a perspective image of meteor shower, leading up to an infinite energy realm.

The infinite swimming pool blurs the boundary between water and land while blending the landscape inside and outside water. The sightseeing in such a journey is not only sports for body, but release of spirit.

The parent-child room is a tree house riotous with color. That is a secret base for Children where to tap their potential.

现代风尚
MODERN STYLE

台湾富邦丰泰接待中心
Reception Center of Rich Country

项目位置：台湾新北
设计公司：域研近相空间设计
设计师：李俊平、康智凯
摄影师：SAM
用材：夹板、瓷漆、卡拉拉白大理石、茶玻、铁件、木皮染色
面积：1 255 m²

Location: Xinbei, Taiwan
Design Company: Inheressence Design Studio
Designer: Li Junping, Kang Zhikai
Photographer: SAM
Materials: Plywood, Porcelain Paint, Marble, Tawny Glass, Ironware, Dyed Veneer
Area: 1,255 m²

富邦丰泰接待中心位于新北市新店山区，附近群山环绕，自然资源非常丰富。春夏时节，山区环境清爽宜人，但秋冬之际山区多雨潮湿的气候形态却又是另一种完全不同的环境条件。除了地处群山环抱的自然环境和反差极大的气候条件特色外，基地内长向坡度将近10米的高低差所呈现出来的起伏地形，对于临时性建筑的阶段性任务来说，实为规划设计上需要面对的重要问题与课题。

本案的规划产品诉求主要以 Richard Meier 等五位纽约著名建筑师群所设计的独栋透天别墅为主。因现行的临时建筑的法规限制，以及基地的水土保持等相关法规，对于临时建筑物的高度及开发面积等诸多限制，所以在销售阶段舍弃了"样品屋"的概念，而是透过接待区的空间氛围去传达以 Richard Meier 为主的白派精神及建筑师想要表达的生活态度。然而基地所处的山坡高低落差极大，为了强调基地所处的自然景观特色也有其视野的方向性限制，加上基地现有的生态虽然杂草丛生，但也留有几株原生的乔木，所以，如何将销售时期所需的接待及洽谈空间合理安排，同时又能表现出日后产品的特色，最后还能同时兼顾生态与环境，将销售时期的环境影响程度降到最低，

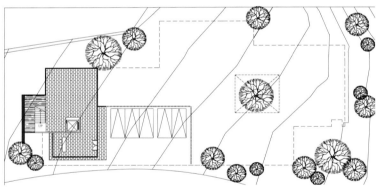
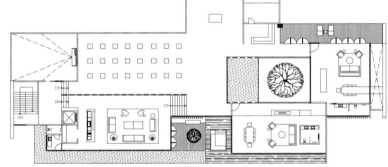

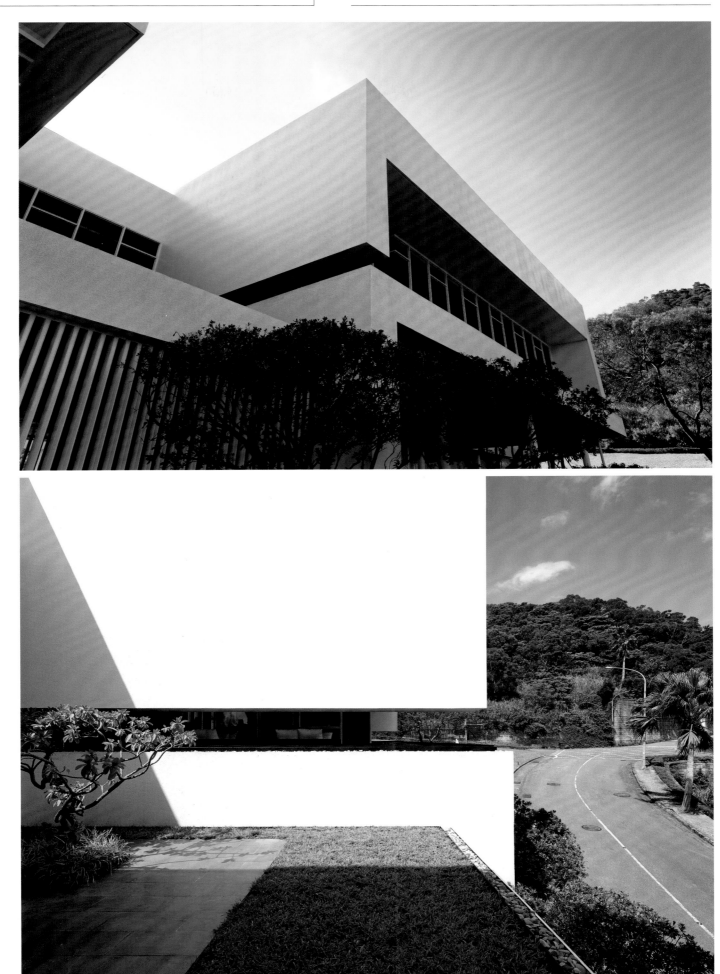

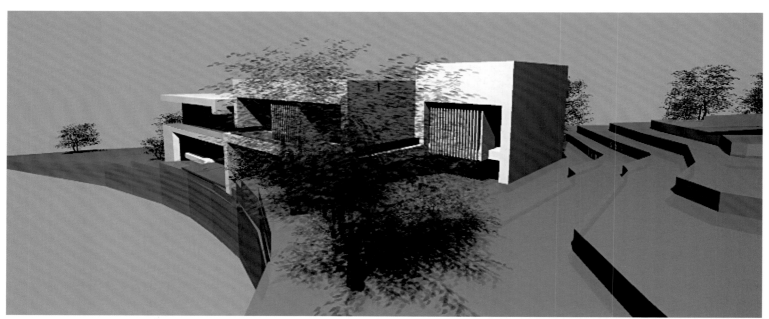

是本案的条件限制，也是最大的期许目标。

本案企图在自然资源极为丰富的台北近郊新店山上，提出一种空间策略，一种既能满足销售行为的需要，又能够尊重自然与环境和谐相处的平衡策略，以及反省空间专业者在这个产业环节里，对于自然环境所能深化的一种态度与坚持。

首先，临时建筑的结构贯穿"杆栏式"的概念，除了对主要落柱进行简易处理外，不仅没有对基地内的山坡进行大规模挖填与整地，反而是顺着坡度进行量体的错层配置，尊重原有的地形地貌与水土保持，将建筑行为对环境的破坏降到最低。其次，所有空间量体避开基地内的原生乔木，透过环抱与延伸的手法，让原生植物与大自然的山景变成空间的主角，透过框景与景深的手法互换主客体，建筑物退居配角，整个环境自然地成了空间的

主角。另外，考量冬季山区湿冷的气候特性，采用大面积的活动开窗，冬天可引进大量的光线与热能，夏天可开启让微气流形成自然的对流通风，减少建筑物对空调的依赖，进而减少能源的消耗。

在空间的安排上，利用山坡高低落差的特性，将接待区、模型区与多媒体区、洽谈区之间互相错层安排，同时创造出开放感与私密性兼具的半穿透空间。主要的接待区向外延伸出铺满草皮的露台与半户外空间，透过绿的延伸将向外的主景视野连接起来。洽谈区以两个隐密性的独立包厢为主，包厢区向内围绕着保留的山樱树，向外则以镜面水池连接向外的开阔视野，在与接待区相邻的侧面则安排了一个降板的半户外区，透过建筑转折的雨遮板围塑出包厢与接待区界面所需的隐私性，同时又能兼顾向外的开放性设计。

在选材上，建筑外墙以夹板与白色磁漆为主，简单的量体与纯粹的颜色传达出光影最原始的表情。室内卡拉拉白大理石在以纯白色为主的基调内，利用人文切割的方式点缀接待区域等重点空间。而染色木地板与木皮的温润质感则有利于缓和纯白过于冷调的空间调性，带出人文的质地。

The reception center enjoys rich and diverse natural resources with hills around. Seasons of spring and summer autumn are easy and comfort for people to live, while the other two seasons of autumn and winter are quite different. For a temporary construction, big climate difference and the high-low difference by the slope almost as long as 10 meters pose challenges.

The reception center for single villas by 5 famous designers from New York, including Richard Meier are subject to laws for temporary construction, laws on water and soil protection, and regulations on height and area of temporary construction. So the concept of show flat has been abandoned, but intended life attitude is expressed in the ambiance. The most challenging challenge is how to give equal consideration to reasonably arrange the sales and the communication spaces, to bring out the future feature of the project and to pay attention both to zoology and environment while minimizing the negative impact on the environment. Every coin has two sides. When landscape is optimized, the orientation of view is limited, just like weeds to go with bushes.

A project puts forward a spatial strategy to meet demands of sales and display in balancing the respect for nature and sustainable development. Meanwhile, reflection by designers on nature with a persistent attitude is revealed in this space. With a concept of balustrade, spilt-level configuration is made along the slope besides necessary column, which with the rejection of digging and filling minimizing the negative effect on the surroundings when holding respect for the

landform and water and soil protection. All spatial volumes are kept away from trees to make the native plants and hills be the hero of the space. With approaches like landscape framing, the building is retreated to be servant while the holistic master. Mobile windows of large area make good response to the wet and cold winter by introducing more lighting and thermal energy while beneficial to ventilation to reduce rely on air conditioner when opened in summer.

The use of height difference fixes the reception area, the model area, the communication area and the multi-function area at different levels, creating a semi-penetrating space both open and private. The main reception area is extended onto the terrace paved with lawn, which links the major landscape by stretching green. The communication area mainly consists of two separate boxes, whose interior is centered on the cherry tree while exterior is oriented beyond with mirror pond. The side adjacent to the reception area is a semi-open area that is lowed deliberately. The rain shelter around the turning point shapes the privacy needed for the box and the reception area, boasting the open design toward the outside.

Plywood and white porcelain paint are more used for the exterior wall. The simple volume and the pure color bring out the most primitive expression of light and shadow. In a setting of pure white, marble diced embellishes key sections. And at the same time, dyed wood flooring and veneer offset the too cold tone by white to embody the cultural texture.

长白山中弘池南区项目售楼中心
The Sales Center of Zhonghong Chinan, Changbai Mountain

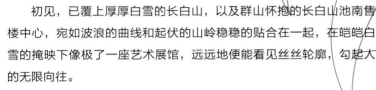

设计公司：本则创意（柏舍励创专属设计）
用材：老松木、木饰面、新月亮古大理石、蒙古黑大理石、黑拉丝不锈钢
面积：1 850 m²

Design Company: Basic Concept (Perceptron)
Materials: Old Pine, Veneer, Marble, Black Draw-Bench Stainless Steel
Area: 1,850 m²

初见，已覆上厚厚白雪的长白山，以及群山怀抱的长白山池南售楼中心，宛如波浪的曲线和起伏的山岭稳稳的贴合在一起，在皑皑白雪的掩映下像极了一座艺术展馆，远远地便能看见丝丝轮廓，勾起人的无限向往。

诚然，设计师在设计此售楼中心的时候对其所处的自然环境以及文化氛围都做足了功课，宏伟壮观的长白山，群山绵延，在此打造休闲度假的多功能养生之所，以崇敬自然的鬼斧神工。

步入神圣的殿堂，远离城市的喧嚣，摆脱世俗的浮华，回归生命纯净的状态，在这里找到一片心灵的净土。一花一木，由自然而生，为自然而用，一呼一吸间，给人最纯净祥和的气息。

设计师独具匠心，没有过多的修饰，整体空间以原木为基础，保留木质最原始的生命力，所用实木材质纹理清晰可见，宛如历史的沉淀，更是自然气质的彰显。在天花、墙面等位置均采用木梁结构，或原生态木材拼接方式，天然去雕饰，线与面本身的体量感和张力便被释放出来。

大堂向阳一侧采用玻璃幕墙，让阳光毫无阻隔地直射进来，窗外山峦叠影重重，那是贴近自然的最佳位置。窗边小憩，与这山水融为一体，是体感的舒适，也是心灵的怡逸。

大自然赋予我们最本质的东西，没有一丝矫揉造作，正如大堂上方的吊灯演绎了小小的宇宙模型。天地万物，不过是一粒尘埃，造物主总是希望我们将这些最纯真质朴的东西呈现出来，不论何种形式，都让人真真切切地感受到生命的力量。

无论是宏观还是微观，设计师都处理得很恰当。在接待室中，设立了木质的屏风，随之穿透的阳光便有了形状和温度，树枝状的台灯和鹿角有序拼凑的吊灯，溪边饮水的麋鹿，配上皮质的椅子，本就是一种艺术的呈现，震撼人心。设计师在本案中强调环保，推崇建筑与自然的融合，在大气的氛围中感受宁静。任由阳光在空间肆意，品一壶清茶，悟一丝禅意。欲在纷纷扰扰的都市中寻找一隅宁静之地，也就非此地莫属了。

The first sight reach the Changbai Mountain that has been coated in snow and the sales center in the arms of hills. A gallery it is like against the snow, an undulating wave that has been plied up tightly with the winding ridges. Its profile in the distance is bound to evoke human desire.

No doubt, without enough a thorough investigation of and insight into the surroundings and the cultural atmosphere, without the sales center. With the backdrop of the grandeur and generosity of Changbai Mountain and other hill, the sales center comes as a multi-functional space for resort, leisure and health care while adoring the uncanny workmanship by nature.

A holy palace here is that's kept far away from the urban bustling and hustling, or the worldly ostentation to return to a pure state of life and to a refuge where to get the pure land of heart, where all plants are growing and used for nature, and where all are destined for the most harmonious sense.

The real ingenuity contributes no surplus embellishment. The basis of log keeps the most primitive vigor and vitality of wood. The primitive wood grain clear and neat is like the sediment by history to embody the natural temperament. Ceiling and walls either employ wood gird structure, or timber splicing. So natural it is to leave out the gingerbread while the force and tension inherent in lines and surfaces is released.

The side of the lobby facing the sun is of glass to maximize the sunlight. Outside the window are layers and layers of hills, a place closest to nature. Resting by the window, you feel fused into the landscape, comfortable in your body and joyful in your heart.

万有引力·售楼部设计
GRAVITY / SALES CENTER DESIGN

What' the most essential for us is from natural without any mincemeat. The chandelier in the lobby accomplishes a small universe, where even the heaven and the land are nothing but particle. No matter what form they appears, the purest and simplest allows us to experience the life force most authentically.

Whether the macroscopic or the microscopic, all is treated properly and duly. In the reception room, there is wooden screen, through which the sunlight comes in different shapes and with various temperature. The dendritic desk lamp and the chandelier orderly pieces with deer horn, the elk drinking in creek, and the leather chair make an artistic presentation altogether, to move people's heart to shake their faith. During the design, the concept of environment protection is stressed more by fusing the construction and the nature for people to feel peace and quiet in the universe. As sunlight coms, a tea in mediation leads to tranquility in the depth of inner heart. And here it is.

保利佛山三山新城西雅图销售中心
Seattle Sales Center, Poly (Foshan, Guangdong)

设计公司：广州道胜装饰设计有限公司
设计师：何永明
摄影师：彭宇宪
用材：新古堡灰大理石、爵士白大理石、木饰面、玫瑰金拉丝面不锈钢、瓷砖、扪皮、夹丝玻璃、GRG石膏天花
面积：610 ㎡

Design Company: Guangzhou Daosheng Decoration Design Co., Ltd.
Designer: He Yongming
Photographer: Peng Xianyu
Materials: Marble, Veneer, Stainless Steel, Porcelain, Fabric, Wired Glass, GRG Plaster Ceiling
Area: 610 ㎡

本楼盘名为西雅图，寓意着充满活力、理想的城市家园。

硬装以条形代码为元素，结合墙身的切割特色，营造一个具有艺术特色的空间环境。在软装上以"心动城市"为主题，延续硬装用绿色座位点缀空间，搭配漫天的树叶，营造四季转换的浪漫意境，像是随着微风旋律舞动，演奏自然的乐章。

书吧区的色彩搭配与组合形式，以及咖啡吧艺术油画、饰品的氛围设计。整个空间通过叶片形态来装饰，让人感受到这是一座有生命的、绿色的生态城市，地毯的形状、书架的形态，都传达这一信息。

这个具有切割转折的特色空间软硬呼应，营造出前卫而又清新自然的空间气氛，让人步入室内就能感受到生机勃勃的气息。

A project of this space that is titled with Seattle to symbolize a urban home abundant in vigor and ideal.

Decoration features in bar-type code, which is fused with the wall-cut traits to create an amenity filled with artistic characteristics, while upholstering takes as its theme "a beckoning city" to continue the use of green seats interspersed throughout, and meanwhile, tree leaves all over creates a romantic prospect where seasons come and go. All together make feelings that here is a natural movement with breezes.

The salient feature of the book bar is the collocation of hue and that of the coffee bar is the atmosphere of artistic oil painting and accessories. The availability of leaf throughout the space allows for feelings here is nothing but a green and vigorous city. The shape of carpet and the form of book shelf is complimentary to convey such a message.

In such a space that is cut brightly and clearly, the air is avant-grade, fresh and natural, where you feel nothing but a vivifying atmosphere.

河南台北晶华接待中心
Taipei Jinghua Reception Cener, Henan

设计公司：大匀国际空间设计
设计师：林宪政
软装设计：上海太舍馆贸易有限公司
用材：深色木饰面、黑色烤漆玻璃、镜面不锈钢镀黑钛、爵士白大理石
面积：800 m²

Design Company: SYMMETRY Design
Designer: Lin Xianzheng
Upholstering Design Company: MoGA Decoration Design
Materials: Veneer, Black Stoving Varnished Glass, Black Titanized Mirror Stainless Sterel, Marble
Area: 800 m²

本项目处于一处四线城市老城区的市中心位置，周边的环境相对破旧与杂乱，正对面是整个城市的一条主干道，车水马龙。建筑外观设计灵感来源于有破洞的折纸，设计师意图通过这种几何形状和现代的手法与周边环境形成鲜明的反差。创造一个强势崛起和破坏性的建筑单体并非设计师的本意，也并非想制造一个冰冷而有距离感的空间。商业与设计意图的融合变成一个重要课题。

在平面设计时，设计师希望在满足所有商业销售机能的同时，通过一些模糊内部空间与外部空间界限的设计手法，穿插水景与光影来制造一个穿透、静谧的空间。同时也想让建筑与外部环境相融合。

折纸形状的白色几何形围合墙体把城市主干道的喧嚣与内部销售空间区隔开来。在墙体上无序分布的穿孔，让阻隔又变得不是那么绝对。

主入口设计师刻意以廊桥的形式架在室外的景观水池上，让人穿过折纸墙体行走在水池上。意图通过这种行走其中的过程，让人体味一种从喧嚣逐渐进入静谧的心理感受。水景上在建筑体与折纸造型墙的夹缝中，松树以对景的形式出现在其中。不管从廊桥看过去或售楼大厅内部通过玻璃幕墙看出来，在白色背景与阳光的烘托下都能感受一种浓浓的哲学意境。

内部接待大厅，设计师用一个与外部对比非常强烈和动态的手法来处理。整个空间由白色完全转变为黑色，墙面深色的木皮和内藏灯带在黑钛不锈钢镜面顶面反射下制造了一个非常虚幻动感的空间。地面黑色、白色石材组合与建筑外墙无序的穿孔造型作了一个呼应一直贯穿整个空间，成为空间的一个主题元素。抛开形体和形式的设计，

通过光影渲染空间氛围是整个设计的灵魂。不管是深色色调的沙盘展示区域与外面白色折纸造型墙体在穿孔与狭缝中透过阳光产生的神秘氛围，还是二楼走廊纯白色空间在阳光下所产生的纯净感，都是在用光影来营造空间氛围，烘托空间主题。同时，不同的反差和对比，流动性和穿透性的手法反复运用、转换,使整个空间变得更加丰富和细腻。

现代风尚
MODERN STYLE

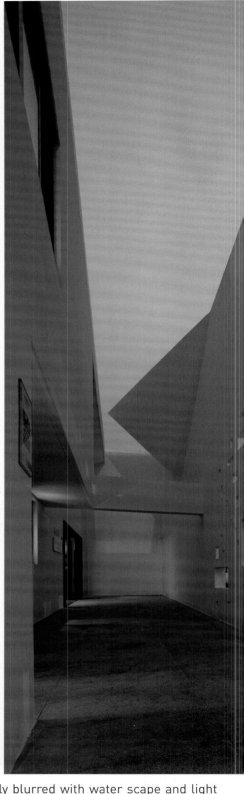

The location in a core area of old quarters of a relatively developed small city is worn-out and in disorder. Opposite to a main stem, the space witnesses heavy traffic on the street. Its design idea comes from paper folding with holes. A sharp but intended contrast with the surroundings turns to very geometric and modern approaches. The creation of a deconstructive single building is not desired, nor a space cold and aloof, far away from its setting. Instead, the fusion of commerce and design aim is our subject.

With all business functions already met, the boundary interior and exterior is deliberately blurred with water scape and light and shadow to pierce into peace and quiet. Meanwhile, the construction is destined to be blended within the outside.

Paper-folding-like geometric shape and walls draw a line between the noisy outside. The holes and cuttings at random at wall makes the separation not so definite. The main entrance purposefully in form of covered bridge allows for a shift for people to feel tranquility in their inner depth of heart when meandering through. Between the construction and the paper-folding wall is the waterscape. The pine modeling

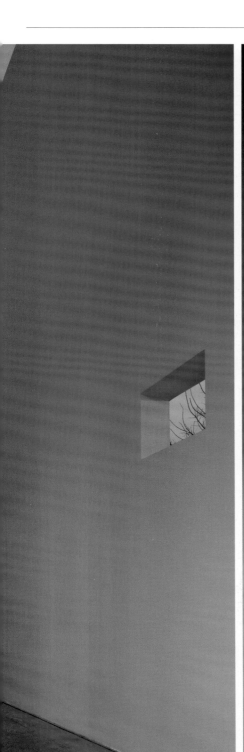

is symmetrical, which can be seen whether from the bride, or in the sales lobby, or through glass wall. Against the white backdrop and the sunlight, there is a strong physiological prospect.

Inside the lobby is the employment that's dynamic and contrasts with the exterior. The holistic has been transferred into black from white. The dark veneer coating the wall and the embedded lighting belt reflect a very illusory world with light from black tetanized stainless steel. The black flooring, the white marble and the unordered openings of the outer wall have been throughout, making a theme element of the project. With the body and model abandoned, the ambience of light and shadow is practically the soul of this space. Both the dark-hued sand table and the white paper-folding wall generates mysterious atmosphere with the aid of sunlight that comes through holes and cracks. Meanwhile, the white of the 2nd-floor corridor brings forward a sense of purity. All are hoped with light and shadow by using and switching contrast, fluidity and penetration time and time again to enrich the space while making it finer and smoother.

郑州绿地中心
Zhengzhou Greenland Center

设计公司：穆哈地设计
设计师：颜呈勋

Design Company: MRT Design
Designer: Yan Chengxun

郑州二七滨湖项目地处郑州市二七新区核心腹地，位于鼎盛大道以南、南四环路以北、大学路两侧区域，是集高层甲级办公、总部办公、商业中心、精品酒店、高端居住于一体的大型城市综合体。

售楼处作为商办类销售中心，出售办公与商业两种空间类型，要求室内设计绚丽时尚，能渲染商业氛围，并具备新地标、国际、品质、现代奢华的效果。所以室内设计风格不拘特定形式，大胆创新出新商业体空间。

室内设计围绕设计灵感，通过墙面、地板、天花不同部位的表现，形成多种立体面的折面效果，在对比的同时又相互搭配映衬，突破原建筑的限定空间。LED灯带在墙面上细细勾勒，意在表现时光交错的质感和现代融合形成的碰撞，又似一种指引，让来到这里的贵宾有深探其由的冲动。

Perched in the hinderland with Dingsheng Avenue in it south, South Fourth Road in its north and Daxue Road on its sides, the lakefront project makes a large urban compound that integrates high-rise first-class office, headquarters, business center, boutique hotel and high-level residence.

Its positon of center of business and office contributes to its interior design being flashy and fashionable, intentionally rendered commercial ambiance when generating global, modern and luxurious effect as a new landmark. So its design is not fixed, but bold and innovative to shape a new business space.

All design is centered on design idea. Expression of different sections of walls and floors allows for stereoscopic multiple folding facets, contrasting and simultaneously collocating to break away the inherent boundary of the building. LED belts on walls are aimed to bring forward a texture of crisscrossing the past, the present and the future a collision between the modern and the contrast, or rather it is stimulating guiding people to explore further.

现代风尚 | MODERN STYLE

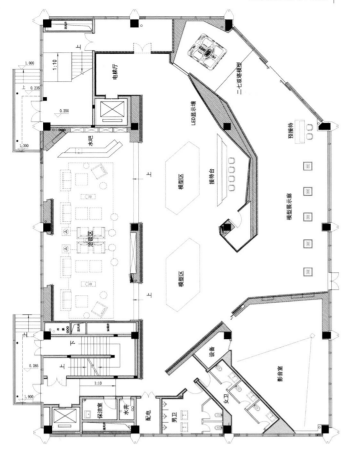
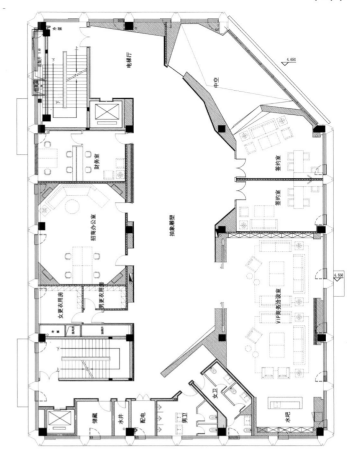

河南平天下接待中心
World-Peace Reception Cener, Henan

设计公司：大匀国际空间设计
设计师：林宪政
软装设计：上海太舍馆贸易有限公司
文案：严旸（协同设计）
面积：1 560 ㎡

Design Company: SYMMETRY Design
Designer: Lin Xianzheng
Upholstering Design Company: MoGA Decoration Design
Text: Yan Yang
Area: 1,560 ㎡

　　本项目的设计需要重新构建新的空间与景观，从百度地图上了解到台北晶华与平天下这两个项目，其实是在城市的同一条主干道上，相距不过六七千米。但由于城市不大，位于建设路与体育路处的台北晶华项目处于老城区的繁华地段，而平天下项目却处在新区与老城区的中间地带，有比较优越的交通条件与生态景观环境。地段的原因，两个项目的定位很清晰，闹市区的项目会偏商业空间，而环境优越的项目自然而然就是居住空间。在一个不大的三四线城市中的同一条路上的两个项目，既要做出差异性，又要体现共同点。当然由于地块大小的原因，这两个项目定位的差异较大，寻求两者之间的联系与共性，反而变成设计师面临的一个重要课题。

　　当初在创造台北晶华项目这个建筑空间时，从最初的想要用体块与水景营造像 KERRY HILL 式的建筑氛围，到为了迎合商业考量再

加入折纸的概念，再到引入一些无序的方块元素来充实空间。最终形成了一个独特的"沉默而稳静的喧嚣"的建筑氛围，我们在不断地加入想要的元素。

而转到平天下项目则是采用流动空间的理念来架构整个建筑空间。用一道道看似动态奔跑的墙体来支撑整个空间结构。同样把景观环境与建筑或室内空间作内外结合，通过模糊内外空间的界限来达到室内室外与景观融合的效果。通过中庭、草坪植栽、水景等一个个的串联，把整个空间融为一体。透过不同的窗口与角度让人与景观有不同的视觉互动与空间体验。景观水池的设置更是再一次试图把"沉默而稳静的喧嚣"理念植入整个建筑氛围中。通过无边景观水体让建筑体有种犹如飘浮在水面上的视觉，想通过这种想象来营造一种状态，营造一艘漂浮在平静湖面的小船，人躺在小船上晒着太阳随风飘荡的状态。主要意图还是想在VIP洽谈室营造一种静谧、懒散而放松的感觉与氛围。

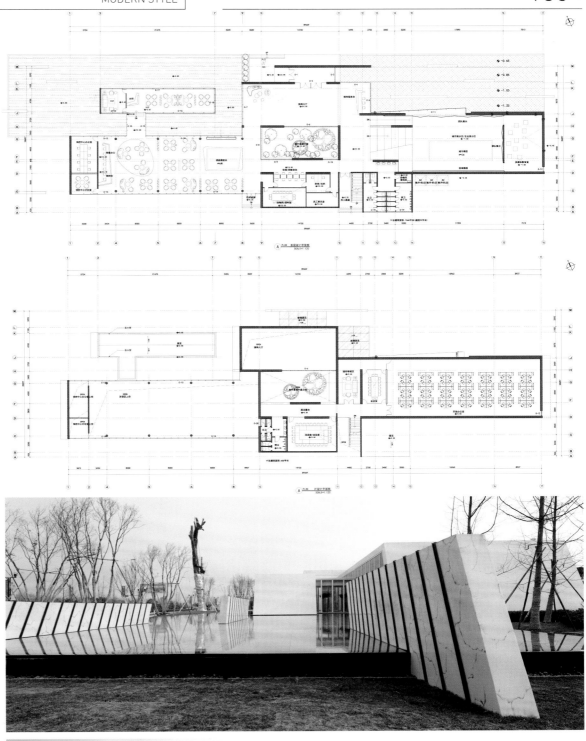

The design for this project needs reconstruction of space and landscape. This one in the middle between old and new quarters and Jinghua Reception Center located in the same street with 6 or 7 kilometers away have to be differentiated. When both are positioned clearly, this one with its more favorable traffic and ecological landscape is mainly for dwelling while that one for commerce. On the contrary, the relation and generality between is another task.

万有引力·售楼部设计
GRAVITY / SALES CENTER DESIGN

现代风尚
MODERN STYLE

As for Jinghua, it's aimed to create a Kerry-Hill building atmosphere with block and waterscape with the use of folding paper to meet commercial function and unordered diamonds to fill in. The final result is the atmosphere, unique, staid and calm. Feelings and elements have been constantly applied and added.

As for this project, it turns to a concept of flowing space to shape the space, where one and after another dynamic wall stand up. The landscape and the interior has been collected to blur the boundary between. The series string of atrium,

lawn, and waterscape joins all together. Different windows and perspectives offer interactions between landscape and people to have spatial experience. The landscape pool is complimentary to the concept silent, staid and steady. With the infinite pool, the building feels as if it were floating like a boat on mirror-like water, where people swing with sun and breeze. That's what's to be acquired, like the tranquility and relaxation in VIP room.

上海虹桥旭辉办公体验馆
The Experience Center of Xuhui Office, Hongqiao, Shanghai

公司名称：上海曼图室内设计有限公司
设计师：冯未墨、弗雷德里克、孙毓婉
摄影师：陈志
用材：有机木、黑钢板、混凝土饰面板、黑钛拉丝、不锈钢
面积：1 500 m²

Design Company: Shanghai M2 Design Limited
Designer: Feng Weimo, Fredric Addey, Sun Yuwan
Photographer: Chen Zhi
Materials: Organic Steel, Black Steel Plate, Concrete Venneer, Wood Flooring, Black Drawn-Bench Stainless Steel
Area: 1,500 m²

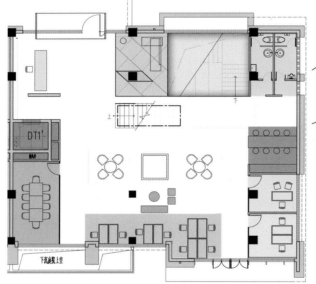
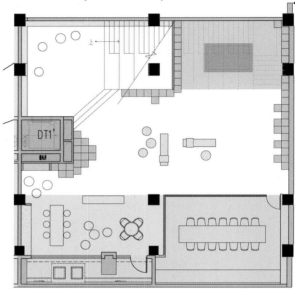

本项目是将独栋办公展示样板间与售楼中心功能相融，创作出独特的体验式售楼处。

本案是一家名为 A+I 建筑设计公司的办公场所，项目共五层，总面积为 1 500 平方米，是一个从建筑构造上去研究终端客户工作方式与环境的独栋办公建筑。

通过打通楼层楼板、加固横梁等手段，来最大限度地引进自然光线，同时得到具有创造性的室内交通动线。景观与城市规划部门、建筑部门及室内设计部门之间的互动方式也将更加的自然轻松。

针对上海市场，更多的考虑环保概念的灵活性，运用再循环木材作为空间核心——悬挑的会议室的表面材料，无论从一层或是三层看都有极好的视野。

整栋办公的定义不仅仅是交流沟通的场所，更像是定制的地下室所具备的概念化娱乐与工作的环境一样，用户可以借助空间工具来体现建筑职业的独特性。

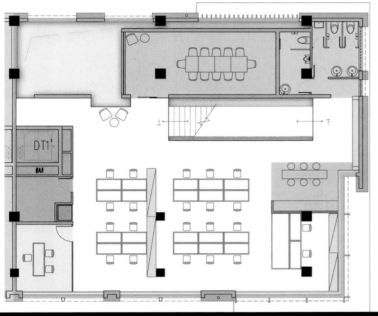
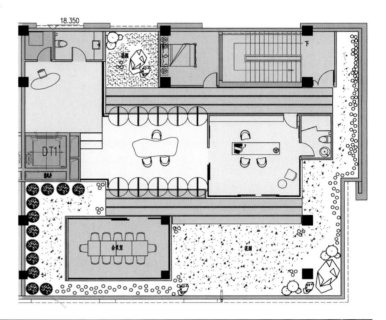

A project this space that fuses the displaying and marketing function of a single building, where to allow for a unique sales experience.
The A+I project is a 5-storey building with an area 1,500 square meters, aimed to make a structural trial to study customers' working and the environment of a freestanding building.
The maximum use of natural light gained by the demolition floor slabs and critical beams is simultaneously friendly to the traffic flow in the interior. The interaction and

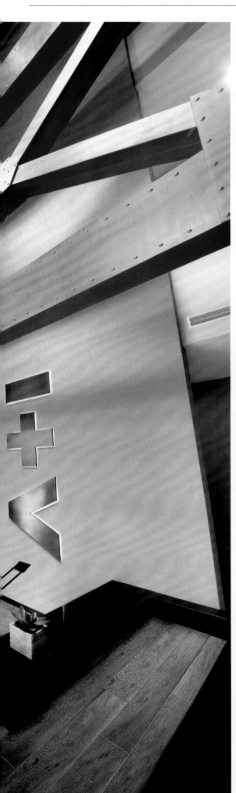

the collaboration among the designing staff, the constructor, the landscape and municipal planning development is more natural and smoother.
The feasibility of environment practice and awareness is Shanghai-localized. The coating material of the conference and one more floor or rather three floors based on its original floors make a signature landmark for vision.
The office throughout the entire building is more a customized basement to hold conceptual entertainment and work than a field focusing on communication, by which the space can be used to embody the unique of construction career.

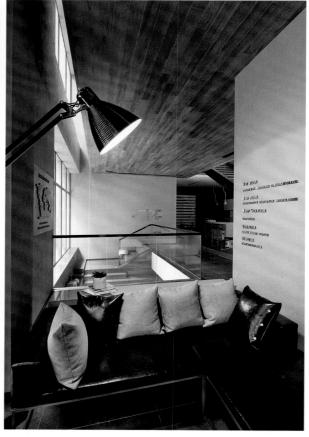

现代风尚 | MODERN STYLE

光明华强文化创意产业园销售展示中心
The Show and Sales Center of Huanqiang Creative Industry Park

项目位置：广东深圳
设计公司：李益中空间设计有限公司
设计师：李益中
用材：爵士白、波斯海浪灰、乔布斯灰、人造石、艺术玻璃、木地板、地毯

Location: Shenzhen, Guangdong
Design Company: Li Yizhong Space Design
Designer: Li Yizhong
Materials: Marble, Glass, Wood Flooring, Carpet

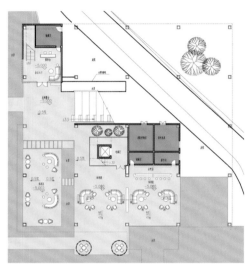
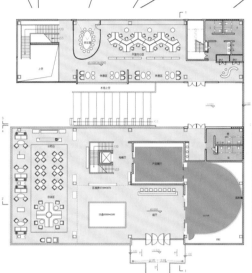

华强文化创意项目位于凤凰城北片区，观光路以北、高铁站以东，总占地面积约 23.09 万平方米，容积率仅为 3.34，总建筑面积约 58.3 万平方米，产业形态涵括研发厂房、写字楼、铂金公寓、潮流商业、3D 展览、云处理等，是光明新区携手华强集团缔造的首个宜居宜业，聚纳生活、工作、休闲、娱乐于一体的大型产城综合体。

而作为其销售展示中心的本案则位于光明新区门户凤凰城，三层高的建筑体量采用低冲击开发理念打造出绝佳的自然环境和舒适的办公空间，给人以绝对的世外桃源般的办公感受。展示中心由两幢建筑组成，通过一条空中廊道串联在一起，构成一个完整连续的空间体系。空间功能分区明显，负一层为吧台休闲区，一、二层为主要功能空间，

包括洽谈区、沙盘展示区、开放办公区、签约区、会议室以及营销办公室等。

休闲区空间开阔，布置虽简单，但颇具情调。简洁现代风格的弧形座椅，简约大方的水吧台，融合在柔和的光线里，营造出一种理想的情调，让来者有温馨、舒适的家的感觉。从中间的楼梯拾级而上，到达一、二层空间。其室内没有繁琐、奢华的装饰，却在简约、低调的气质中渗透着与众不同的品质，精致大方，色调沉稳，高雅而不张扬。项目展示厅运用了3D全息投影技术，将装饰性和实用性融为一体，形式新颖，科技感浓郁，给观者以全新的感受。而不同的灯饰运用不仅界定空间范围，与艺术品、花饰等一起配合使用更进一步美化空间与强化空间气氛，增强空间的优雅气氛和文化气质，使空间更为动人。

Located in the north quarter of Phoenix Town, the Huanqiang Creative Industry in north of Guanguang Road, and east of high speed rail station, covers an area of 23.09 thousand square meters, where the plot ratio is 3.34. In the building area of 58.3 square meters, the space can be used for factory, office, platinum apartment, leading business, 3D exhibition, and cloud processing. As the first residence by the local area and Huanqiang Group, it makes a compound for living, working, and having leisure and entertainment.

And the show and sales center in the gateway of the local area, the 3-floor building is developed with minimized effect on its surroundings to present a superexcellent and cozy office environment to allow for working experience

only available in a land of idyllic beauty. Linked with the two buildings is a gallery in the air to make a continuous system. The spatial function is clearly cut, the basement for office and leisure, the 1st and 2nd floors as main part to accommodate the communication area, the sand table, the open office, the contract-signing area, the conference and the marking office.

As for the leisure area, it's open and wide, and quite stylish, though equipped simple. Into the soft light are the arched chairs of modern style and the water bar that's

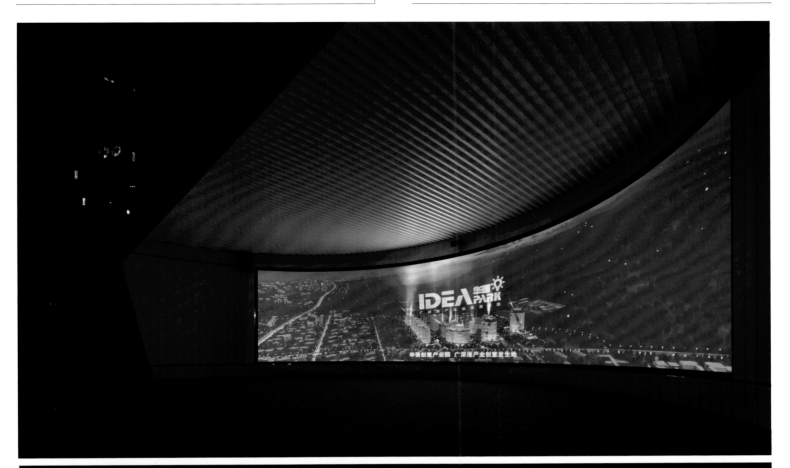

simple and yet grand, from which flows out an ideal emotional appeal for visitors to feel themselves at home. The stairs in the middle lead upward into the 1st and 2nd floors, where there is no complicate or luxurious decoration, but different quality that's delicate and magnificent, refined and fairly reserved with the simple and conservative temperament. With 3D projector technology, the decoration and the practicality is fused together, innovative in form and abundant in science and technology to offer all-newly experience. Differentiated lighting has not only defined the spatial scope, but further beautifies and reinforces the beauty with art pieces and floriation. Consequently, when its grace and culture is enhanced, the space is becoming more vivid.

珠江科技数码城销售中心
Pearl River Science and Technology Digital City Sales Center

项目位置：广东佛山
设计公司：广州共生形态工程设计有限公司
设计总监：彭征
参与设计：谢泽坤
面积：1 500 ㎡

Location: Foshan, Guangdong
Design Company: C & C Design
Director Designer: Peng Zheng
Participant: Xie Zekun
Area: 1,500 ㎡

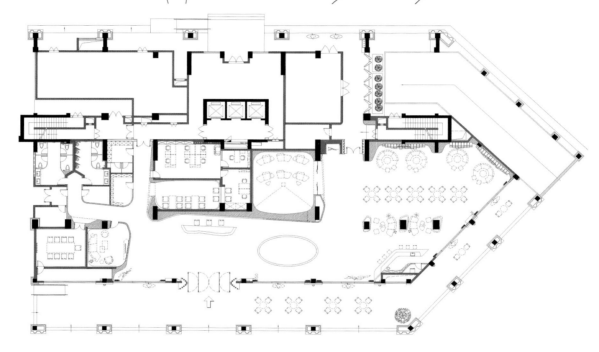

本项目是一个位于古典风格建筑里的售楼中心，当我们在城市生活太久时，容易感觉麻木和忽略自然风光所带来的最初感动。快节奏的信息时代，也经常让人们忽略日常生活之美。设计师突破建筑的限制，思考着自由、流动空间的可能性。繁星、山峰、河流、雪花或是甜甜的巧克力都曾是我们儿时的感动，他们以抽象的形态提醒人们居住环境的重要性。

A project this space that is perched onto a classical sales center. Exposed to city for too long, people are quite easy to feel numb and ignore the initial touch by landscape. As it is, the fast-speed information age contributes to people's unawareness of their daily life. And this project is aimed to be a reminder of the importance of dwelling in its abstract sense by breaking away the construction limitation and reflecting on the potential of a free and flowing space, where these touching items in childhood are readily available, like stars, peaks, rivers, snow drops, and chocolate.

福州信通售楼中心
Xin Tong Sales Center, Fuzhou

设计公司：大阅艺术设计机构
设计师：何承春、何金华
用材：铁片、实木地板、钢化玻璃、茶色不锈钢
面积：400 ㎡

Design Company: Dayue Design Ltd.
Designer: He Chengchun, He Jinhua
Materials: Iron Piece, Solid Wood Flooring, Toughed Glass, Tawny Stainless Steel
Area: 400 ㎡

本案室内空间设计简洁有力，没有过多的装饰，同时却处处散发着柔美和优雅，这就是本案设计的最大特色。

设计师运用条形不规则的金属装饰线条，在一虚一实之间刻画出空间中的光影变化与节奏，大大增强了仪式感和层与层之间的互动性，让这个面积本来较小的售楼中心更加通透。利落的线条勾勒出空间的构架，简单而大气。深色调的木纹搭配金属，朴实无华的色彩融合，不会惊艳，亦不浮夸，却让人在无限的回味中感受到速度、优雅和愉悦。

Here features no complicated decoration. Everything is succinct, of great force, and morbidezza and graceful. That's the most salient feature of this project. With irregular metal decorative lines, change and rhyme of light and shadow has been mapped out in a real-virtual approach, greatly enhancing the ritual sense and the interaction among layers. So the smaller sales center is more transparent. With neat lines, the spatial structure is simple and yet of grandeur and generosity. The match of dark-hued wood grain and the metal, and that of plain colors are anything but amazing or showy, for people to feel the speed, the grace and the pleasure with aftertaste.

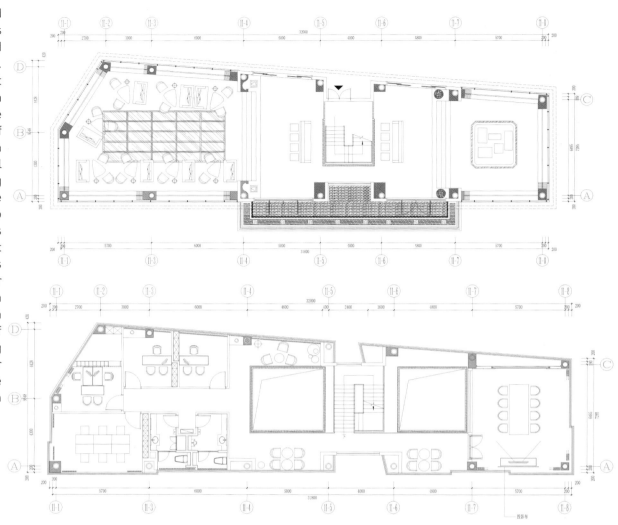

台湾青田青接待会馆
Green in Green Reception Center

设计公司：域研近相空间设计　　Design Company: Inheressence Design Studio
设计师：李俊平、康智凯　　　　Designer: Li Junping, Kang Zhikai

青田青接待会馆位于台北市大安区和平东路一段196号，该会馆由域研近相操刀设计，摒弃冗杂的点缀，从最原始、最质朴的建筑原点、材料原点、创意原点出发，创造出一个含蓄、和谐、具有丰富文化内涵和充满人文气息的舒适空间，让身处其中的人们，与时光相拥，静享细水长流。

接待会馆极为独特的外观造型与立面设计，使之成了该地段让人驻足欣赏的一道美丽的风景线。圆润灵动的线条与不规则的界面组成的奇形怪状，容易激起人们的好奇心，驱使他们继续往里一探究竟。进入室内，设计通过独具创意的隔断划分方式来布局不同的功能区域，而木格栅与玻璃镜面隔断的使用则显得尤为突出，使得两个空间产生似虚而实的视觉对接。

再细看，空间中遍布充满质朴古意并带有禅风的木质媒材，从顶面、墙面、地板，到家具，其透露出来的文化底蕴与艺术气息，沁人心脾，衍生出一种安宁的心境。设计采用天然石材点睛，温润中又有着无限的安定。而各式大小不一的艺术品，也为创造这充满人文风格的环境氛围发挥了重要的作用。当夜幕降临，灯光洒落，那袅袅升腾的缥缈与诗意，暗香浮动，把一个韵味空间辉映得更加充满魅力。

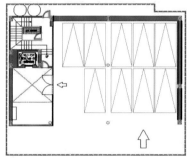

现代风尚
MODERN STYLE

Located in Daan District, Taipei, Green in Green Reception Center makes a space connotative, harmonious and rich in culture and humanity by employing the most primitive and plainest construction feature and material, and applying innovation, where people can indulge themselves in peace and quiet, while unware that time passes by.
Its appearance and facade make it a beautiful scenery line for people to come to a stop. The grotesque in shape of mellow and full lines and irregular interfaces is so easy to stimulate people's curiosity to step into to explore. Inside is the partition creative and innovative to define different functional spaces. Meanwhile, the wood grating and the glass are so outstanding, spaces separated with which are actually divided but visually joined together.
Onto the ceiling, the wall, the flooring, and the furniture, is the applied wood medial. It's simple and unadorned,

and of Zen and antique taste to generate a heart-relaxing setting. With texture of nature marble, the design is warm and definitely tranquil. Art pieces big or small, play a pivotal part in shaping the cultural ambience. As the dark sets in, light comes misty and poetic with secret fragrance to compliment the spatial charm and taste.

厦门宝龙一城售楼部
Bao Long One Mall Sales Center

项目位置：福建厦门
开发商：宝龙地产
设计公司：KLID 达观建筑工程事务所
设计师：凌子达、杨家玛
面积：1 500 ㎡

Location: Xiamen, Fujian
Developer: Bao Long Real Estate
Design Company: Kris Lin Interior Design
Designer: Lin Zida, Yang Jiayu
Area: 1,500 ㎡

　　本项目设计的两个主题为"相对"与"融合"，并探讨如何从相对发展到融合及两者之间的关系。

　　"相对"：宇宙间的一切事物都存在着"对立"与"融合"的两面，是相互对立的而又相互和谐存在的关系，这是物质世界的定律。

　　"融合"：把两个对立的元素透过相互交叉与卡接的方式合而为一，形成一个完整的个体，达到了"融合"。

　　"室内功能"：最后完成的"融合体"，即建筑同时也反映出室内空间的三个主要功能区——展示、销售、招商，实现了建筑与室内的完美结合。

Bao Long One Mall Sales Center is a project that has two themes, "opposite" and "fusion", by which to explore the relations between.

"Opposite" means the two sides of opposite and fusion, a dipole universal in university. That is the law existing in the material world, of being opposite while fused together.

"Fusion" is to combine two opposite elements by being intercrossed and connected into an integrated unite to reach a harmonious state.

"Functions in the interior" is an impeccable mixture of displaying, marketing and investment attracting, three different functions under the same roof.

现代风尚
MODERN STYLE

万科未来城接待中心
The Reception Center of Future Town, Vanke

设计公司：大勺国际设计
设计师：林宪政

Design Company: SYMMETRY Design
Designer: Lin Xianzheng

万科未来城接待中心位于杭州，由大勺国际设计操刀设计，通过独特而鲜明的空间构成手法来表达对空间分割的理解，使空间利用最大化。1 022平方米的大空间，采用干净简单的块面与线条进行处理，同时材质与细节装饰，简约、艺术、实用，希望给每一位访客带来一种不一样的视觉体验与空间感受。

一层由接待区、3D多媒体影音区、模型区、洽谈区、水吧台以及儿童活动区组成，二楼包含办公室、会议室、接待室及等候区等，各个功能区各自独立成景又景景相融。进入大厅，设计让运动器材成为个性别致的装饰品，两辆白色脚踏车离开了地面，被悬挂在天花板上，让人有着电影情节般的狂想。再加上其经过变幻的美丽造型，增加了空间的艺术感。再往里走，极具特色的顶面造型吸引着我们的目光，许许多多排列整齐的白色管状物从天花板上倒挂下来，夸张且美轮美奂的艺术造型，让访客的感官和心灵受到了一次洗礼与震撼。白色六边形墙面搭配黑色大理石地板，白色的扶手搭配白色阶梯，黑白之间的自由转换带给人们澄净分明的感受，整个空间显得简洁大气，时尚优雅。

洽谈区不规则的沙发与茶几，儿童活动区的娱乐设施，六边形蜂窝墙面，几何形拼贴地毯，灯具……这些极具视觉冲击力的几何体，外形美观别致，时尚简约，在不经意间传达着接待中心的蓬勃动感与不竭的能量转换。

色调的运用以黑色、灰色、白色为主，并在其中加入一些鲜明的色彩，如柠檬黄、新绿、橘色等，高调炫目，增加了空间的色泽饱和度，在感官上给人一种强烈的刺激，这也是设计上对色彩处理的点睛之笔。

Located in Hangzhou, the reception center of Future Town, Vanke, is intersected spatially with distinctive approaches to maximize the use of area. All 1,022 square meters is endowed with clear and crisp surfaces, blocks and lines. Materials and decorations are brief, artistic and practical to allow for unique visual and spatial experience.

The first floor consists of reception area, 3D audio-vison room, sand table, communication room, water bar and area for children activity, while the second floor houses office, conference room, another reception room and waiting room. All functional rooms are kept individualized and interspersed. The lobby features in sports facilities, like two white bicycles

万有引力·售楼部设计
GRAVITY / SALES CENTER DESIGN

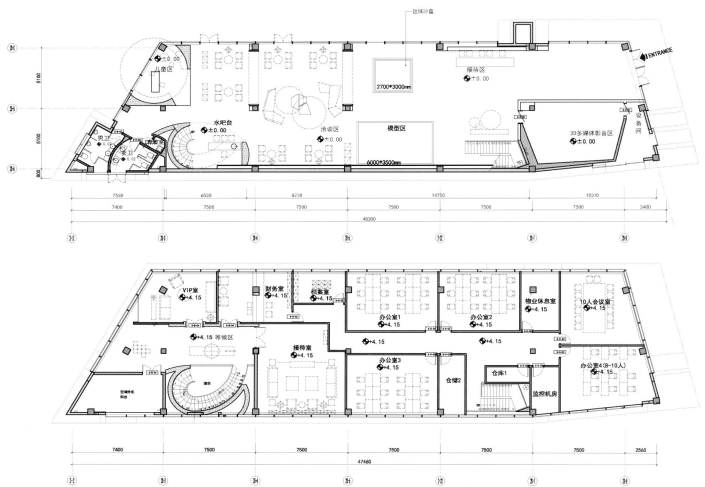

down the ceiling which to generate film-plot chimera, and whose changeable beautiful modeling is complimentary to the art sense in the space. The further top surface of the ceiling is so special and eye-catching, where rows of white pipes hang upside down offer a baptism on organs and spirits. The white hexagon wall with the black marble flooring, and the white handrails with steps of the same hue bring out a sense of purity, the space thus becoming succinct and magnificent when of fashion and grace.

The irregular-shaped sofa and the tea table in the communication area, the recreational facilities in the area for children activity, the carpet in geometric shape and the lighting altogether exert a strong effect on vision. Beautiful, fashionable and simple in appearance, all articles convey vigor and vitality, and inexhaustible energy conversion that is taking place in this project.

The dominance of black, gray and white, is interspersed with bright colors, like lemon yellow, verdancy and orange. The glitz of lofty tone enhances the spatial saturability in keeping off the dim feelings to allow for a strong stimulus on sense organs. This makes a finishing touch in terms of color treatment.

台湾温布敦19接待中心
Wimbledon19 Reception Center, Taiwan

设计公司：域研近相空间设计　　Design Company: Inheressence Design Studio
设计师：李俊平、康智凯　　　　Designer: Li Junping, Kang Zhikai

基地位处台北市内湖区民权东路六段，正面对着世界大学运动会网球中心预定用地。广告销售企划也紧抓住这个概念，朝一个运动休闲宅的方向包装。接待中心的外观以网球的特性与身体的律动韵律交织发展出独特的线条语汇，企图呈现出一种力量与优雅同时并存的美学形式，来凸显基地的魅力与独特性。

The location in Nei Lake District, Taipei. It is opposite to the place reserved for tennis match World University Games. Its advertisement and sales makes the best use of such a concept, oriented toward sports and leisure. Its appearance features in the traits of tennis and the swimming body, intentionally to represent an aesthetics of both force and grace to protrude the unique and special of the geographical existence.

重庆旭辉乐活城体验中心
The Experience Center of LOHAS Town, Chongqing

公司名称：上海曼图室内设计有限公司
设计师：冯未墨、弗雷德里克、喻晋
摄影师：陈志
用材：锈钢板、水泥纤维板、木地板、灰钛不锈钢
面积：1 380 ㎡

Design Company: Shanghai M2 Design Limited
Designer: Feng Weimo, Fredric Addey, Yu Jin
Photographer: Chen Zhi
Materials: Rustic Steel Plate, Cement Fiber Board, Wood Flooring, Gray Titanized Stainless Steel
Area: 1,380 ㎡

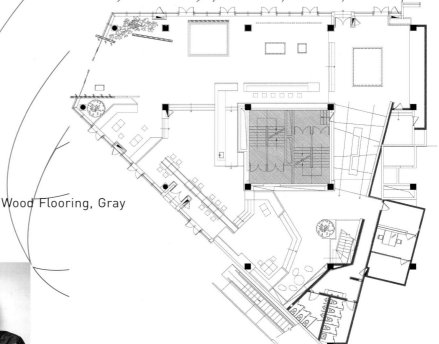

通过对现代售楼功能需求进行深入分析，打造了一个咖啡主题的体验式售楼空间，满足售楼功能的同时降低售楼空间给客户带来的压迫感。

整个空间采用生态自然的材质——木地板、水泥纤维板、锈钢板打造一个简洁、现代的空间。在模型区通往洽谈区的位置设置了绿植长廊，而洽谈区大量书柜及壁炉的设计，使人有一种舒适自由的洽谈感受。

完整的售楼动线提供一个流畅的参观旅程，空间跌级的变化以及空中的连桥设计予人深刻的印象，同时跌级的空间感受也契合了重庆的城市特征。因建筑二层空间有挑檐设计，致使空间的挑空区域设置在靠核心筒区域，而二层沿幕墙区域则设计为多个独立的VIP空间，使得室内空间产生一种小中见大的效果。

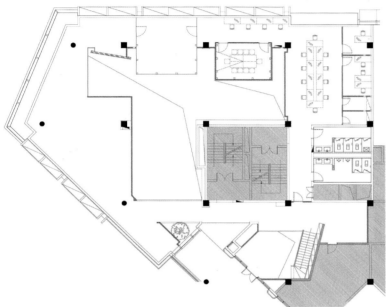

A project this space is that has undertaken deep analysis of modern sales center, which now has taken coffee as its theme in meeting sales function and reducing the pressure onto customers.

Throughout is the employment of ecological materials to create a concise and modern space, like wood flooring, cement fiber board, and rustic stainless steel. Between the model and the communication area lies a greenery corridor. The latter space allows for comfortable and free experience, embellished amounts of book cabinet and fresco.

Along the complete traffic line comes a flowing visiting journey. The changes of levels and the bridge strike an imposing impression. When the level change well fits in with the local feature of Chongqing, the hollowed-out pace is fixed near the core tube thanks to the cornice of the two floors. Flanked with the glass wall are freestanding VIP rooms, which generates an effect that there is much in little.

宁波华侨城欢乐海岸售楼处
The Sales Center of the Happy Coast, OCT, Ningbo

设计公司：壹舍室内设计（上海）有限公司
设计师：方磊、朱庆龙、柴润知
用材：哑面白洞石、黄铜饰面、橡木、透明玻璃、夹丝玻璃
面积：1 000 m²

Design Company: One House Inerior Design
Construction Company: Fang Lei, Zhu Qinglong, Chai Runzhi
Materials: Marble, Bronze Veneer, Oak, Transparent Glass, Wire Glass
Area: 1,000 m²

本案位于宁波，地处奉化江河流转角处，地理位置非常优越。是当地最具影响力的开发项目之一，有宁波最大的滨江主题公园、亚洲最大的音乐喷泉广场、国家艺术馆、主题酒店、商业、住宅等综合体项目。设计师接受委托设计一期住宅销售售楼处地块，即安排三栋样板展示区和一栋主题售楼处。

经过现场考察，虽有奉化江水系景观但现场已被工地覆盖，基地现有景观不理想。设计师努力说服甲方把售楼处南向市政绿化带进行景观改造，变成一个对市民开放的主题景观公园。这个公园是城市的延伸空间也是售楼处的延伸空间，这个区域让建筑和城市街道完美结合。同时也是映射整个地块项目的核心精神：打造一个欢乐的海洋生活体验场所。

首先将地块南北分成3个区域，即南向为公共景观区域，北面为3栋样板展示屋，中间部分就被围合成一个"核"，这个"核"就是销售展示中心。

房屋销售本身并不只是销售被钢筋水泥围合的数据，而应该给客人

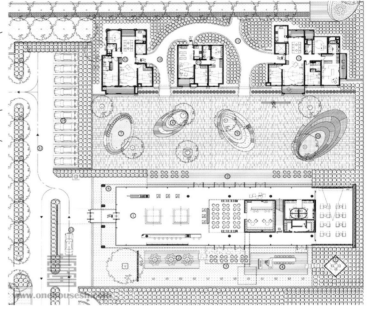

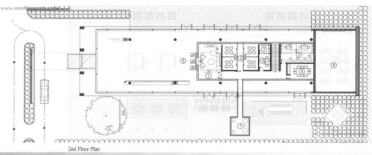

2nd Floor Plan

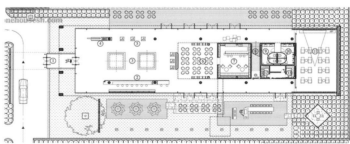

呈现未来的环境和居住形式。在生态日渐恶化的今天，激发客人购买欲望的应该是良好的环境和优质的服务。所以一开始设计师就决定要把周边的环境最大化的呈现出来，而售楼处则是一个承载环境的载体。具体的做法如下。

1. 为售楼处围合一个绿色的人造景观区域。

2. 为售楼处地块注入一池水。"水"是宁波欢乐海岸的主题，也是整个项目的灵魂。把"水"这个主题通过售楼处进行延续，水像一个随风变化的舞台，可以把周边的景观映射出来，同时也映射着建筑的内部。水的出现模糊了建筑内外的界限让建筑与环境充分融合。

3. 在水上加入平台和柱网，在四周围合透明玻璃形成室内空间。底层4.5米宽的透明玻璃可感知四季的变化并把户外的绿色尽收眼底。此外，还在户外的水域加入平台形成户外休息区域。

4. 为幕墙上部加入一层金属表皮，如建筑的皮肤。随着光线的移动给室内带来不断变化的光影。

5. 为建筑外部插入两个量体。一个为入口，另一个是空中的露台。入口的雨披是从建筑内部生长出来并且延伸至户外的石材百叶墙，构成了立体的围合空间，为售楼处提供一个安静尊贵的入口。而另一个户外的空中露台则是从建筑内部空间直接穿过建筑表皮延伸至户外，形成一个三维立体的视觉景观，为二楼的VIP室和签约室提供一个良好的户外休息空间。

The project is located at Ningbo, a city in the corner of Fenghua River. The section where it is has the largest theme park in the local place, the largest music fountain square in Asia, national gallery of arts, theme hotels, and other commercial projects. The project includes three show areas and one building as sales center.

As the landscape of the local water system is damaged during construction, the owner is persuaded to retreat the south part of the site into the municipal green belt as a landscape park open to citizens. The theme park is considered as extension of both the city and the sales center, which well connects the building and street. Moreover, the landscape park reflects the core sprite of this area, a place to offer happy ocean experience.

Firstly, the plot of land is divided three parts from north to south: the south is for public landscape, the north 3 show flats and the middle core as the sales and display center.

House selling is not only to witness marketing database that takes place in a concrete setting, but also to provide a future-oriented lifestyle and dwelling environment to customer. Healthy environment and quality services are factors that motivate people to own properties as pollution becomes one of the greatest social issues today. So the presentation of the surroundings of this project is aimed to be maximized. And the sales center makes no doubt a carrier of the environment. As below are the solutions:

1. Build landscape to enclose sales center.

2. Build pool around the sales center. "Water" is a significant element of Ningbo Happy Coast and the soul of this project. Such a theme realizes its extension with the aid of the sales center, for water can reflect both the landscape and the interior, as a stage that can change with breeze. Its appearance blurs the boundary between the interior and the exterior, so the architecture and the environment can be well fused by water.

3. Build platform for outdoor recreation and column grid into water. With glass to embrace the interior. Through the width of 4.5-meter glass comes into the seasonal landscape.

4. Attach metal surface to glass, like the skin of the building. As light comes and go, shadow within changes constantly.

5. Plug two parts into the building, one being the entrance and the other alfresco terrace. The poncho seems to grow from the interior into the louver wall, making an embracing space within three dimensions and providing a peaceful and dignified entrance. With the terrace that cuts through the façade into the exterior, a three-dimensional visual landscape comes into existence, a good resting place for the VIP room and the contract-signing room on the 2nd floor.

台湾三本千晴接待中心
San Ben Qian Qing Reception Center

设计公司：域研近相空间设计
设计师：李俊平、康智凯

Design Company: Inheressence Design Studio
Designer: Li Junping, Kang Zhikai

将更好的环境与建筑结合，给人类带来更好的生活，是建筑人与设计师的使命。三本千晴接待中心位于桃园近郊的一块大草坪上，附近自然生态资源丰富，到处弥漫着绿茵芳草的清新气息。结合基地的地理优势，设计师把"让室外的大自然与室内和谐地融为一体"的设计理念运用到接待中心的规划，展现出热爱地球、拥抱自然、珍惜资源三大精神，让人类、建筑、环境相存相依，和谐共生。

除了尽量保持基地最大的绿覆率外，建筑物以一个有机体的概念出发，设计出独特的造型，使之可以随着自然与四季变换而有不同的生长形态出现。为了将绿意延续到空间内部，设计师打造了玻璃幕墙，让光线穿透性进入，不仅让人从里面就可看到室外的盎然绿意，光线洒落在各种温润的家具上，还使空间流露出一种生活的气息与纯正的味道，淳朴而雅致。内部设计以木材和石材混合打造，温润的木色，玉石感斑驳的石纹，每一个细节都隐隐地流露出温婉的人文气息。家具结合原色的木材和灰色的软坐垫进行组合设计，软硬兼施，设计感十足，流露出一派温情。

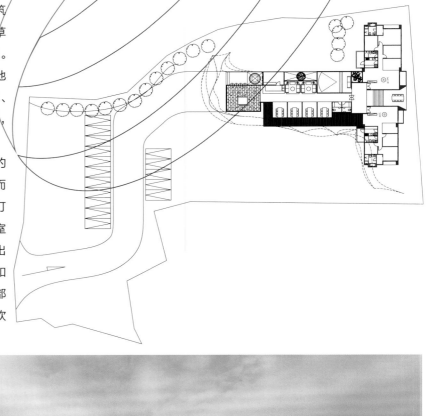

To combine better environment and building to give people a better life is the mission for architect and designers. This is a project in a large lawn in outskirt of Taoyuan, a place rich in ecological resource and fresh fragrance. In blending the local geographical advantage, designers have skillfully joined the harmony of its interior and exterior into the design to bring out three spirits to love the earth, to hug the nature and to treasure the sources. So man, building and environment can coexist in a peaceful state.

Besides the maximized forestation rate of the base, designers

start from an organic concept to design a unique shape which can have varying growing forms with season change. In order to continue the external greenery, glass curtain is used not only for daylight to come, but adds a life sense and an authentic flavor. This is simple and refined. The interior is of wood and marble. With the wood warmth and the jade-like vein of the marble, each and every detail confides in a mild cultural ambiance. Out of the wood of the furniture whose color is kept original and the cushion of gray, flows out a strong sense of design as well as warmth.

周浦绿地缤纷广场售楼处
Sales Center Riotous Square, Greenland (Zhoupu, Shanghai)

设计公司：集艾室内设计（上海）有限公司
摄影：三像摄 / 张静

Design Company: G & A Design International
Photography: Threeimages / Zhang Jing

　　周浦绿地缤纷广场售楼处位于上海市周浦公园旁，唯一的高级商业，以"城市中的花园"为主题打造最舒适的花园商业。售楼处延续着整洁的主题，以"花园中的精品店"贯穿整个空间，展现出品质与生态的完美结合。

　　整个空间以弧线贯穿虚实结合，变化组合。一推开门，映入眼帘的便是闪耀精致的树阴隔断，以叶为元素，组合变化，陈列延展，配以镜面与光线，呈现出极具记忆点的接待空间。伴随着梦幻的光影进入中庭，融模型区、展示区、洽谈区为一体，充分的联动形成了最佳的营销模式。弧线延展，以椭圆的模型台为池，鱼群式吊灯似鸟般盘旋。晶莹剔透，洽谈区软装以极富品质感的驼色为基，点缀亮丽的蓝色、橙色，焕发出不一样的光彩。如湖面荡漾般的地毯、生动地划分空间。弧墙上精心挑选的饰品以橱窗模式展现，更为精致的形象渲染着精品的商业氛围。

　　空间的最大亮点便是背倚窗外美景，面向静水面的室内花园，包含着VIP与高级洽谈区功能。幽静的氛围令人不禁停下脚步，并不由自主地引入其中，心生向往，因水而生，沿水而居。以发光膜为顶膜，模仿天光，精心设计的图像与植物色彩配比，令植生墙更具艺术感。水面中倒影清晰可见，不经意中便会发现由青蛙泛起的阵阵涟漪（生态驱虫）。坐在与水连成一片的VIP区中，享受着别样的尊贵与静谧，欣赏着整个空间的静与动，感受着花园与精品氛围的二重奏。

Nearby Zhoupu Park, the project of this space makes the only high-ranking commercial space with a theme of "a park in city" to make the most comfortable garden business space. The sales center employs a succinct pattern with boutique in garden throughout the whole space to bring out a perfect combination of quality and ecology.

The holistic area undertakes a fusion solid and virtual. With the door, a view of tree shadow as partition comes into view. Leaves are arranged in change, accompanied with mirror and light to bring out a reception space that you would put into your memory. With light and shadow, you come into the atrium, a collection of model, exhibition and communication, whose linkage makes the optimal marketing mode. The circle model table looks like a pond, with chandelier of fish group circling like birds. The communication area, clear and transparent, is coated in light tan with blue and orange to flash different. Looking like ripples of a lake, the carpet divides the space. The careful selection of accessories on the arched wall seem to have been put into window with their delicate image to render the boutique ambience.

The internal garden against the landscape and waterscape outside makes the brightest spot, where to house VIP and senior communication. The peace and quiet brings steps to an end. Desire to enter and then settle down is stimulated because of water. Lighting film is used onto the ceiling to make daylighting image. Graphics deliberately designed make a good match with the plant color, so the vertical garden is planted with more artistic sense. Peeping into the reflection onto the water, suddenly you perceive ripples by frogs, but Nearby Zhoupu Park, the project of this space makes

现代风尚
MODERN STYLE

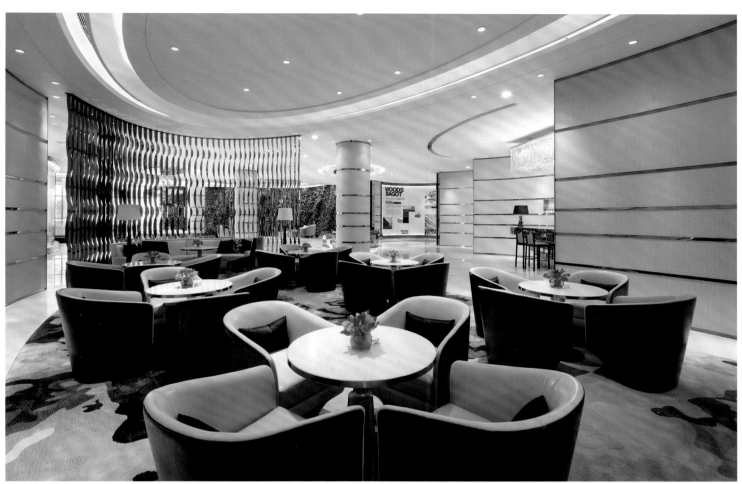

the only high-ranking commercial space with a theme of "a park in city" to make the most comfortable garden business space. The sales center employs a succinct pattern with boutique in garden throughout the whole space to bring out a perfect combination of quality and ecology.

乡林山海汇
Collection of Forest, Hill and Sea

设计公司：郑唐皇计划设计室	Design Company: Zheng Tanghuang Planning Design Studio
设计师：郑唐皇、陈健国、黄国玉	Designer: Zheng Tanghuang, Chen Jianguo, Huang Guoyu
摄影师：李国伟	Photographer: Li Guowei
用材：钢构、抿石子、白橡木、贝壳杉、瑞士檀木、壁布、清玻璃	Materials: Steel Structure, Pebble, Oak, Shell Fir, Sandalwood, Wallpaper, Clear Glass
面积：2 005 m²	Area: 2,005 m²

基地隐身于台北淡水八势路旁的红树林半山巷弄内，前临红树林湿地保护区，后倚大屯山系万顷碧绿，在结合地形与原始林相的环境条件下，尽可能地保留原生森林，试图以借景与造景的方式，从环境观察中找出建筑表演的可能性。

因此，建筑依着缓坡顺势而生，把机能空间抬高于基地之上，留下一楼的开放空间，以亲切的姿态迎接参观者，从基地的景观规划延伸到周边绿意，充满着独有氛围的地区特色，从建筑外观到内部机能，让到访者亲自去感受人与环境间的互动，曲径通幽、拾绿而进，以极简、平和、自然的布局，描绘出让心灵休憩的居所。

由于基地邻山环绕、前方为淡水出海口、观音山脉，设计师提出把接待中心外观塑成大地的剖面图为建筑架构的概念主轴，随着人的观看高度的改变，可以看见利用木头来表示与自然相近第一进的语汇，大致量体偏向邻近的大屯山峦翠，而朝向开放的树林环境其表现形式则以白色建筑板框勾勒出转折线条，从地面翻转至墙面并向天花板伸展延续，刻挖的墙洞通往下方的样品屋区，同时也突显了墙面厚度。再者，挑高的建筑量体使一楼成为可遮雨的出入空间，入口处以三角形挑高与二楼形成串联，刻意放大和突显的自然场域，让通往二楼的动线就像在绿坡拾级而上一般，约十米挑高的迎宾玄关。

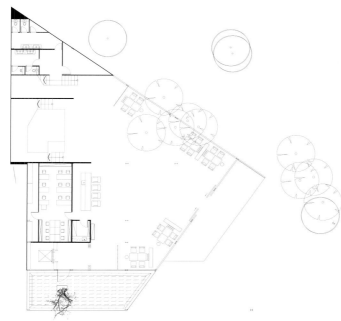
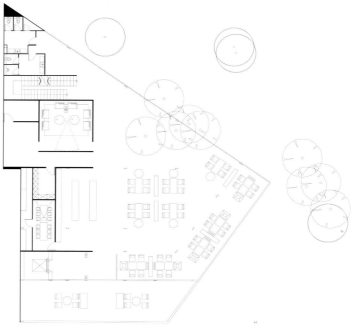

　　随着架高的阶梯转折，开始布局所需的机能区域。进入室内后，主空间迎向树林的开放面，运用水平穿透的视觉和大面积的落地窗营造了流动的空间感，拥有三个向度的清玻璃纳入了更多的阳光和绿意，让室内外更好地连接起来，而围塑出来的内景，随着人的移动来感受在不同楼层都能被环境包围的感官体验。内部树木凌驾于楼板之上，与建筑的钢构支撑不谋而合。原始林的树干同时也消弥了既有的柱体结构，呈现了树林意象的有趣画面，露台水光也越过窗线进入室内，透过窗前镶嵌的玻璃地面可见水光粼粼，伴随着水样窗景创造不同层次的真实平面。

　　动线转至屋顶层，望向观音夕照水池如镜映出浮云山色，设计师期盼来到这里的访客除了看见饱满绿意，更可以借由这样的写意空间，充分体验宽敞的空间视野和环境尺度，领略半山居所带来的崭新风貌。

Reclusive in a lane of red woods, this project really enjoys a good geographical location with a wetland reserve area in front and green hill at back. The combination with the terrain and the primitive forest that is maximized in this project is aimed to find more possibility of construction out of its surroundings by landscape borrowing and making.

Along the gentle slope, functional areas are elevated higher above the base, with the ground floor open and amiable to welcome visitors. Landscape plan is extended around with the unique regional feature filled here and there. Visitors can feel the interaction from the building façade to the internal function. The winding path leading into the depth to draw a prelude of a minimalist, peaceful and natural layout for creating a refuge where to relax mind, refresh soul and rejuvenate heart.

The appearance of earth's section as the conceptual axis is a good correspondence to its neighboring physical hills and river mouth into the sea. Views change as perspective varies: wood serves as the debut to get close to nature, the volume is oriented toward the ill and white board and frame opposite to the forest sketches turning lines that extend from the flooring and stops onto the ceiling via the wall. The openings on the wall lead to the show flat downward, setting off the thickness of the wall. The raised high volume shifts the ground floor into a shelter from rain. The elevated-high triangle at the entrance makes cascade with the second floor. With the field purposefully expanded, the traffic line to the second floor seems to be winding upward across steps. The reception hall is as high as about 10 meters, accessible to a world to accommodate functions.

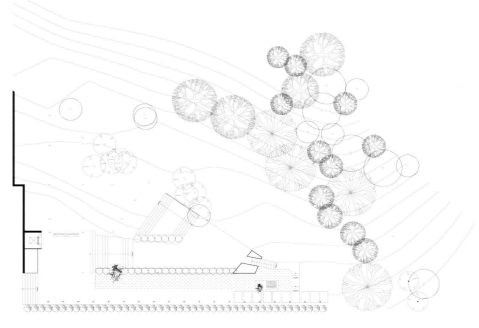

The main space in the interior faces the forest, where horizontal-penetrating vision and landing windows boast the spatial dynamics. Clear glass links the interior and the exterior by taking in more sunlight and green. People within thereby feel as if they were in different environment when on every floor. Trees in the interior rise above the floor slab, echoing with the supporting steel structure. When offsetting the original pillars, trunks maintained present a vivid scene of woods and forest. Waves across terrace comes into the interior with glass flooring by the window helpful to create more layers.

With the evening glow, the pond on the top floor reflects the hills and the clouds. Here offers not only green, but experience of a large spatial view in a dwelling setting halfway up the hill.

B | Oriental Legend

东方传奇

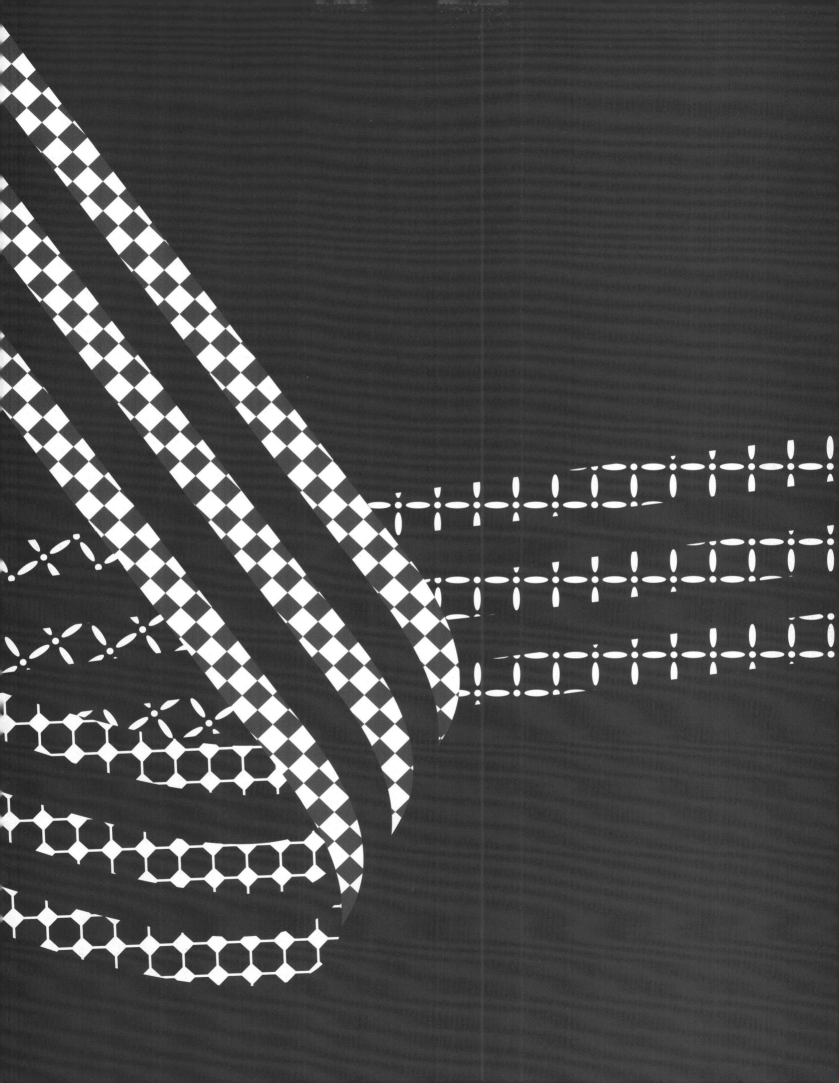

上海中山润园售楼处
Run Garden Sales Center, Shanghai

设计师：连自成
参与设计：曹重华、孙杰
摄影师：张嗣叶
用材：硬木、胡桃木、水云石大理石、
灰木纹大理石、黑钛不锈钢、灰镜
面积：935 m²

Designer: Lian Zicheng
Participant: Cao Chonghua, Sun Jie
Photographer: Zhang Siye
Materials: Hardwood, Walnut, Marble,
Black Titanized Stainless Steel, Gray Mirror
Area: 935 m²

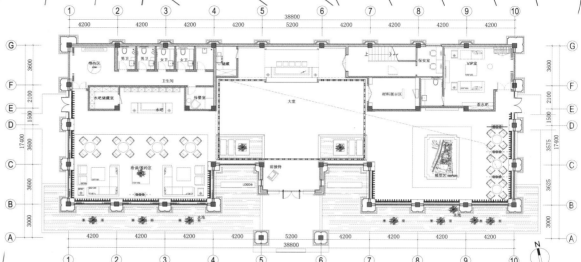

"竹林下，小溪旁，遥望山峦叠翠，抬头即蓝天"这样桃花源般的景致，可以说是都市人对于闲适生活的全部梦想。而今，连自成将这一段静好岁月的自在带到了上海这个不折不扣的繁华之都。

中山润园售楼处地处上海凯旋北路，紧邻苏州河畔，是长宁的中心区域。千年的历史文化造就了苏州独特的性格，而连自成也巧妙的将这古风古韵，以及其洒脱的姿态隐于中山润园售楼处935平方米的空间内。

其实对于一个设计师而言，营造一个三维立体的空间并不难，难的是在空间内注入生活，即我们常说的"人情味"。正如售楼处大厅的空间设计，进入之后有一种置身于鸟笼的视觉想象，它是对蝈蝈笼原型的再创造，也是整个空间里"最生活"的一部分。熟悉中国历史的人大都对这一物件不陌生，在古代它常出现在达官贵人手中，是尊贵和休闲的标志。将它设立在售楼处最显而易见的地方，旨在传递一种情愫：以鸟笼之名致敬历史，令中国风得以具象体现的同时，也让"偷

得浮生半日闲"的惬意弥漫至整个空间。

设计不仅要情理之中还要出其不意,在整个售楼处的设计中,到处弥漫着写意自然的气质,但仔细观察不难发现,细节之处的考究才更能突出其尊贵奢华的本质。悬挂于大厅中央的"万重山"由25万颗璀璨的水晶组成,是6名经验丰富的技师耗时两个月而成。它以连绵山峰的造型出现,并且和门口的桃花相搭配,不仅映衬了传统中国风的主题,也在一定程度上彰显了售楼处大气磅礴之势。

"我寻求的是一种平衡,写意自然与精品高档、金碧辉煌与粗衣麻布的平衡。"设计师连自成说。

在此空间中，从微观世界看自然的景象，中国写意的山水意境全盘托出，这也是设计师所表达的场所精神。从饰品的选择到细节的控制，每一步都极为考究。因为市场上的手法难以满足连自成对于空间的期望，因此这里的装饰品全部私人定制，用水云石的线条及质感表现东方美学的意境；将健康环保的 PVC 材质织造成的大地色地毯，再配以木皮色家具，传递着浓郁的中式风情。不仅如此，景泰蓝、铜器、荷兰青花瓷、Tom Dixon 错落有致的点缀于空间，也让传统与现代不期而遇，共同演绎出美轮美奂的视觉效果。

Scenery only available in the Peach Garden has always been all for people in pursuit of leisure in urban city. Bamboo forest, creek, greenery hills and blue sky comes to Shanghai, a veritably busy capital.

Perched onto Kaixuan North Road, and adjacent to Suzhou River is the sales center in the core area of Changning District. The special characteristics of Suzhou that has been developed for hundreds of years has been transferred into this space of 935 square meters, vintage, free and easy.

The creation of a 3D space for a designer is not hard. What's difficult is to inject life within. That is human touch. The lobby allows for feelings, that you feel as if you were in a bird cage, recreation of katydid nestle which is one part to most embody daily life. Historically, bird cage used to be frequented in the hands of prominent officials and eminent personages, as a sign of dignity and leisure. Its existence in the most seeable place is destined to convey a kind of sentiment: to pay homage to history, to specify the Chinese style and to overspread throughout that time can be spared to spend on leisure off a busy schedule.

Any design should both be reasonable and exert a surprising effect on its viewers. Everywhere is an enjoyable image rooted in nature. The detailed significance set off the essence of the exalted luxury. The overlapped hill hanging in the middle consist of 250 thousand crystal pieces, done with 6 experienced technicians in 2 months. A modeling appears in form of unbroken hills while making matching with the peach blossom, not only setting off the theme of Chinese style, but highlighting the great momentum of the sales center in some extent.

"What I seek after is balance, one between enjoyable nature, boutiques resplendent magnificence and coarse clothing and linen", says Lian Zicheng.

In this space, natural scenes can be perceivable from microscope perspective, where the image of landscape all over China is presented overall. That's the spiritual field intended by the designer. All from the choice of accessory to the detail are paid significant attention. Due to the shortage in the market that can reach the designer's expectation, all embellishment pieces are bespoke, like the lines and texture of the marble to the expression of oriental aesthetics, the earth-colored carpet of environment-friendly pvc and the veneer-wrapping furniture to flow out the bountiful Chinese style. Meanwhile, cloisonné, copper ware, blue and white porcelain from Holland and Tom Dixon are well-proportioned here and there, with which the modern and the tradition meet by chance. A visual effect comes beautiful and gay.

苏州建发地产中泱天成项目售楼处
The Sales Center of Zhongyang Tiancheng, Suzhou Jianfa Real Estate

设计公司：深圳市昊泽空间设计有限公司
设计师：韩松
用材：尼斯木饰面、米黄石材、柚木
面积：550 ㎡

Design Company: Horizon Space Design
Designer: Han Song
Materials: Veneer, Marble, Teak
Area: 550 ㎡

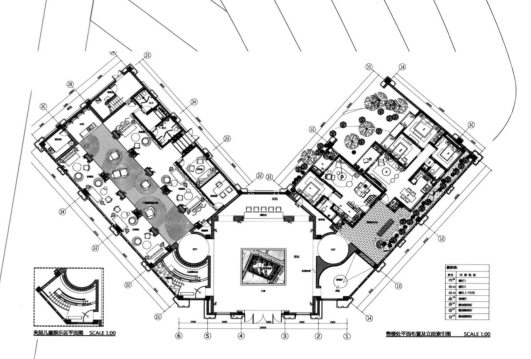

无论个人还是人类的发展都会经历两个过程。第一个过程是人性对动物性的超越，即文明、社会、规则、安全；第二个过程是对人性的超越，往往体现为宗教或哲学上的形而上，或终极的神性。一般理解为精神上人性束缚的自由和解放。而在当下社会的极速发展和变革中，我们每个人都无一幸免地时时刻刻经受着对人生意义的纠结和拷问。普遍的精神困局来自于无法对人生现实目的性的超越，即超越功利、欲望、知识等一切的束缚。

碰巧读了"庄生梦蝶"的小故事，会意于庄周竟用如此浪漫诗意的智慧追求自由。虽然充满悲剧性的惆怅，但也让人读来神清气爽，满怀希望。作为设计师，我们常常会体悟到语言文字对人智性的表达是有很多的障碍，而视觉表达作为一种语境，往往能摆脱这种困境。正好也借用这个小故事的灵感，让每一位来访的体验者都能有各自不同的愉悦和放松。当然我们也奢想而不敢妄言，能引起暂时的精神脱轨，思想的自由……如能此，我们的努力将善莫大焉。

每个人都在追求人生的答案。每当读到下面这段文字，心中便充满了透彻与感动。

"南有悬樋，以承清水；近有林，以拾薪材，无不怡然自得。山故名音羽，落叶埋径，茂林深谷，西向晴空，如观西方净土。春观藤花，恰似天上紫云。夏闻郭公，死时引吾往生。秋听秋蝉，道尽世间悲苦。冬眺白雪，积后消逝，如我心罪障。"——《方丈记》

Man has two insurmountable phases whether as individuals or as social group, when they try to achieve development. The first is the surpassing from animality to humanity. That is civilization, society, rule and safety while the second the surpassing from humanity, which is frequently embodied with metaphysics in religious or physical level, or ultimate divinity. Personally, the restrained humanity has been freed and liberated. In a context of dramatic development and changes, no one can escape by luck from the torture and reflection where we can be in the universe. A process it is that comes really hard to get access to

because of personal obstinacy. The story of Chuang-Tzu Dreams Butterfly run into leads to my understanding how intelligent Chuang-Tzu is to seek personal freedom with such poet romantics. Though disconsolate for its tragic token, it is refreshing and full of hopes. As designers, we often sense the obstacle for writing to express human wisdom, a problem that can be solved if visual carrier can be turned to. And so in this project, the inspiration from this story offers visitors joy and relaxation hard to explain but differentiated in each heart. We certainly aim not to crystalize spiritual rejuvenation even temporary and free thought, but at least that's what we are trying.

Everyone is in the process of pursuing life answer. Whenever reading the following in Reclusion and Poetry by Kamo no Chomei, a Japanese monkey (1155-1216), I'm touched and open to transparency.

In order to have clear water, a barrel is suspended in the south; right there for firewood collecting, a forest is readily accessible. A hill used to be named music feather, where fallen trees have covered paths, forest is prosperous while valley is deep. Sight into the west reaches a clear sky, like the pure land in the west. The flowers on vine in spring are like pink clouds in the sky. In summer, it said that King Guo can lead me into the Salvation Boulevard. In autumn, the harvest fly tells of all sorrow and sadness a living person has to experience. In winter, with the disappearance of white snow goes away the sin in my heart.

扬州湖滨名都销售中心
Lakefront Fame Capital Sales Center, Yangzhou

设计公司：上海无相室内设计工程有限公司
设计师：王兵、徐洁芳
摄影：三像摄 / 张静
用材：柚木、铁刀木、发光云石、米兰灰、爵士白石材、皮革、真丝壁纸、布艺软包、金属镜面
面积：1 500 m²

Design Company: Shanghai Wuxiang Interiors
Designer: Wang Bing, Xu Jiefang
Photography: Threeimages / Zhang Jing
Materials: Teak, Indian Rose Chestnut, Marble, Leather, Silk Wallpaper, Fabric Upholstering, Metal Mirror
Area: 1,500 m²

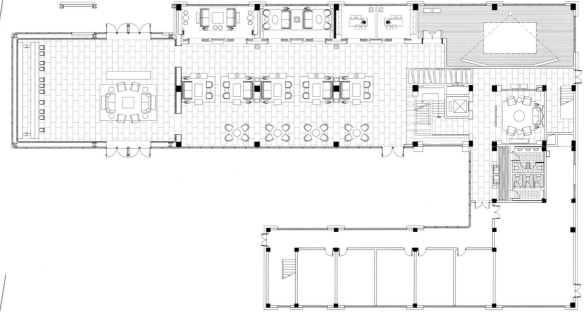

家,一个宁静放松的空间,有一点惊喜,有一丝依恋,这是设计师想要传达给每一个来访客人的第一感受。入口巨大的装饰壁炉,让每个初访者都带着一丝好奇和惊喜,开始细细品味。密集的纵向格栅,恰到好处的留白使得整个挑空大堂在保持神秘之余让室内外景观有了对话的可能,让来访的宾客也入了"画",成了"景"。棕色系的色彩搭配,空间上显得沉静高远。户外阳光明媚,室内光影交织,让人有一种置身丛林,神清气爽的感觉。

运用对称的两个门洞将两个形态不同的空间连接。洽谈区所传达的舒适放松的感觉,犹如在丛林里找到一片开阔地,纵向连续的装饰隔断让空间有序展开并且延伸,分合有致。舒适的新中式家具散落期间,案上,展开的书卷,袅绕的青烟穿透整个空间,坐下来品一壶清茶,细细欣赏窗外的景致,好一个"书香门第",美哉!

洽谈室被规划在北侧一个个空间的"洞"里,设计师想借助立体构成的手法,在空间中切割有序的"洞",如同一个个被切开的宝盒露出内胆,显得神秘、精致。精心挑选的米驼色编织皮革与金属结合,纵横的分割使得皮革包块立刻有了手工定制的感觉。同时,设计师还将天花灯具、空调风口等构件与造型巧妙结合,隐蔽、协调却不失功能。墙面上,满墙的大书架,是用现代的手法演绎东方传统美学,浓浓的书卷气息感染每一个人。

在建筑的核心地带,巧妙地穿插进一池"清泉",让室外的景延伸到室内,精神、气息、神韵以及各个方面都能相互传达。远远看去像是飘在空中又像隐在地里,几尾红鲤,生机勃勃。

整个设计采用大地色系的色调,运用温润的材质,呈现简洁典雅的空间,建筑、景观、室内三者互相穿插,相互渗透。

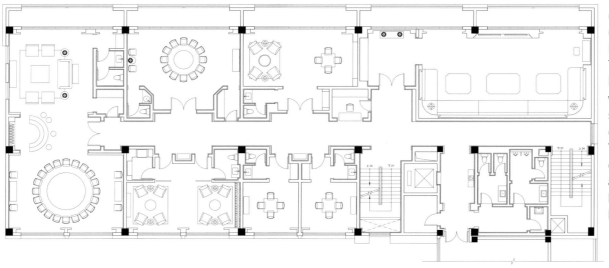

Home is actually a space wh
to enjoy peace and relaxat
with kind of surprise a
feelings hesitate to leave. Th
the primary experience an
home space is intended to e
immediately to the viewers.
decorative fresco stands hug
the entrance, offering curio
and marvel to each who t
taste what comes to his eyes
at leisure. The dense vert
grating and the duly area
blank starts the potentia
have a dialogue between
interior and the exterior w
keeping the hollowed-out lo
in mystery. Visitors feel noth
but come into the pict
as part. Against the bro
backdrop, the space looks l
staid and calm. When it's su
outside, the interior is fly
light and shadow, where you
refreshed as if you were expo
to a jungle.

Two symmetrical door openi
link two different areas.
communication area is easy
comfortable, where you se
to have got access to a w
land in a jungle, the lengthw
decorative segment extends
space into order, the furnit
of neo-Chinese style is rea
available here and there, an
the table with folded book, a
in green smoke contributes
good view oriented beyond. W
a family of scholars here is.
Reasonably landing into
empty boxes in the north
the communication roo
With 3D approaches, orde
openings are like the li
of already cut-open jewe
box, mysterious and delica
The carefully-selected knit
light tan leather is combi
with the metal. The leat
seems to be customized o
cut vertically and horizonta
The skillful combination of
ceiling lighting, the outle
air conditioning is hidden
harmonious. The whole-w
book shelf presents the tradi

of oriental aesthetics with modern approaches, from which is the strong volume sense to affect anyone here.
In the core area lies a pool, which takes in the landscape, so spirit, breath and charm are communicating mutually. Looked far away, it's like floating in the air and buried into the land. Red corps are injecting vigor and vitality.
In a setting against the earth hue with warm material to present a succinct and elegant space, the building, the landscape and the interior have been interspersed and interwoven.

保利·阳江银滩N2售楼中心
N2 Sales Center, Poly

业主：保利（海陵岛）房地产开发有限公司
设计公司：广州道胜装饰设计有限公司
设计师：何永明
参与设计：道胜设计团队
摄影师：彭宇宪
用材：新古堡灰大理石、黑洞大理石、花岗岩大理石、黑白根大理石、白色人造石、巴洛克金大理石、文化砖、木饰面、生态木、玫瑰金拉丝面不锈钢、黑镜钢拉丝面、夹丝玻璃、铝镀黄铜管
面积：2 280 ㎡

Client: Poly (Hailing Island) Real Estate Company
Design Company: Guangzhou Daosheng Decoration Design Co., Ltd.
Designer: He Yongming
Participant: Daosheng Design
Photographer: Peng Yuxian
Materials: Marble, Cultural Brick, Veneer, Greener Wood, Stainless Steel, Wired Glass, Brass Planting Tube
Area: 2,280 ㎡

设计重点

项目采用中国传统园林以及日本枯山水的表现手法，将整个空间打造成度假休闲、高端有品质感的售楼中心。

生态木的质朴结合大理石的刚毅，使画面大气之余更显端重，在整个沉稳的气氛中处处流露自然的气息，灵动的游鱼为空间平添几分雅趣。

设计概念

海，深邃庄重。上至天灵，下至心魂。以宽为广，以沉默为情，以深沉为重，以气势为力。本项目位于阳江银滩，四周环海的条件让整个设计将海的元素以及灵魂延伸到整个室内空间。

整体空间色调沉稳，浅蓝色家具搭配硬装的暖灰色调，

使整个空间氛围舒适而宁静。略带中式韵味的家具与饰品，更突出空间的独特品位。

天花上的吊灯好似欢快的鱼儿吐出的一串串泡泡，为空间平添几分雅趣，也让人仿佛置身在宽阔的大海，得到身与心的放松。

在过道中，运用中式传统园林的手法，把单一的大空间分隔成若干个小空间，相互连绵、延伸。用石子搭配木条，创造出自然休闲的视觉效果，简洁的落地雕塑在打破空间沉闷的同时，与周围气氛相得益彰。

Design Feature
With the technique of expression of Chinese gardening and Japanese rock garden, the whole space is shaped into a high-level and qualified sales center of resort and leisure.

The simple and unadorned fused with the masculine of marble makes the space further solemn when magnificent. Against the staid throughout is the natural with swimming fish adding something graceful and refined.

Design Philosophy
Deep and grave, Sea can reach the heaven as well as the depth of heart. When going as widely as possible, its quiet implies its passion. When sinking where there is no access to get, its momentum has its power and force. And this project is located on Silver Beach, Yangjiang. It physical location with sea around finds its way within by employing sea's element and soul.

The whole hue is staid and steady, where the baby blue furniture goes together with the gray decoration, the holistic space thus becoming comfortable and calm. Furniture and accessories of Chinse style more or less highlight the unique of the space.

Chandeliers are like bubble by fish, swimming here and there to add more elegance and grace, where you feel as if you were in sea with heart refreshed and body rejuvenated.

Treated with Chinse gardening approach, the large aisle is cut into smaller ones, each extended into another. The pebble with the batten sets up a visual effect of leisure and nature. When breaking apart the stiff and dull inherent in the space, the concise landing sculpture and the ambience around is brought out the best in each other.

台湾玺悦会所
Seal-Pleasure Club

项目公司：天坊室内计划
设计师：张清平
面积：11 273 m²

Design Company: Tianfun Interior
Designer: Zhang Qingping
Area: 11,273 m²

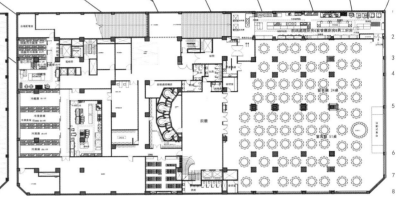

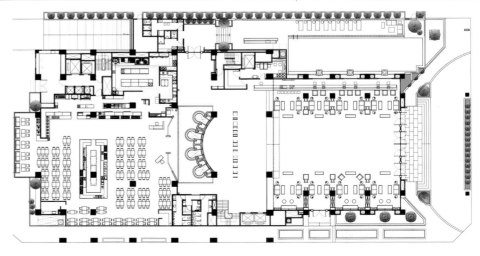

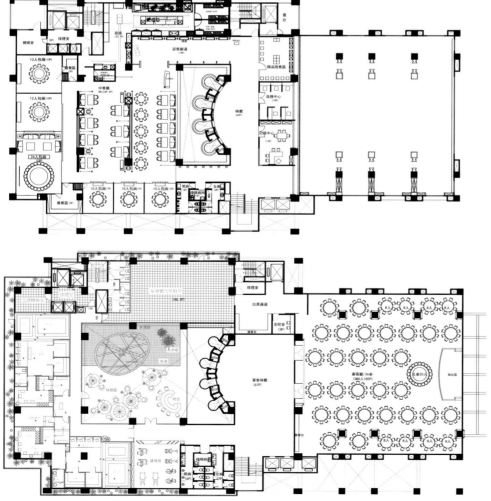

本案整体规划设计概念新颖独到，不仅展现设计的清丽优雅，更同时为将来管理执行的透明与完整，作出完美方案，同时处处落实以客为尊的经营理念，作为本次整体设计的核心思维。

酒店的客人，能体验到的，不再是传统豪华酒店随处可见的常规作法与通用元素的使用。取而代之的是一种处处流露出低调奢华与典雅的当代元素，将宛如艺术品般的空间，与所有的设施完美融合，诠释出度假休闲环境中，强调抚慰身心的细节，让客人在空间与服务上都能获得无与伦比的精致体验。

让客人满足自在的需求，创造更胜于家的体验。整座建筑的设计目标是从里到外，优雅地与自然环境和谐共生，遵循自然的力量，让人类的精神得以升华。从里到外给人一种低调奢华而宁静平和之感。即使再浮躁的心，来到这里也会得到抚慰。

Innovative and unique in its overall design, the project not only exhibits an elegant and beautiful design, but exert a sustainable development for its future use. Such a perfect plan bears in mind that the business is guest-oriented, making the core design for this project.

What's to be experience is not what can be available readily and ordinary elements in traditional luxury hotels, but a low-key luxury and graceful contemporary elements everywhere, which shifts the space into an art piece, where all facilities have been fused in an impeccable state to interpret a leisure and resort environment to soothe and calm disturbed heart and tired body, so guests get nothing but an incomparable experience.

The salient feature to reach guests is the experience better than that at home. The ultimate aim of the design is to find a coexistence between nature and the space external and internal. When the law of nature is followed, human spirit is sublimed. A reserved but sumptuous peace and quiet is oozed from inside to outside. No matter how blundering it is, your heart can be comforted.

C | Foreign Country Sentiment
异域风情

台湾国王城堡会所
King Castle Chamber

项目公司：天坊室内计划
设计师：张清平
面积：5 305 m²

Design Company: Tianfun Interior
Designer: Zhang Qingping
Area: 5,305 m²

国王城堡会所位于台湾高雄，建筑外观强调巴洛克奔放的建筑诗学，恢宏的感觉不言而喻。在一楼的规划上，以宴会厅、泳池两个功能区域，作为主要规划空间，并搭配了其他次要的功能，包括餐厅区域、儿童户外游戏区。酒吧 Lounge Bar 是一个过渡空间，是大厅与泳池之间的穿廊，融入了少许现代感的造型。天花板的装饰与水晶灯的垂坠感，使人一改庄严隆重的意象，而以放松风尚取而代之。

度假设施的重点空间——户外泳池，为了让酒吧 Lounge Bar、餐厅等室内空间可以与户外泳池连成一气，特别采用高达 6~7 米的长式玻璃落地窗，借由大面自然采光，增强户内与户外的通透感。

至于地下一层的公设空间，首重功能性，规划俱乐部、宴客、家庭聚会、视讯会议等空间，满足企业主以及家族交际宴会的需求。第二门厅则采用华丽的岗石建材，迎宾车道上有一面纯粹大理石材质制作的雕塑艺术墙，让宾客在第一眼即留下气派辉煌的印象。

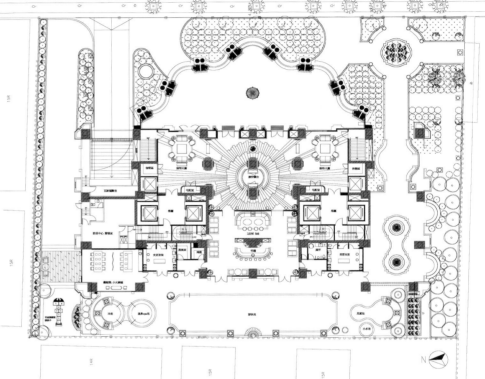

Located in Kaohsiung, the project of King Castle Chamber stresses the bold and unrestrained poetics of Baroque construction. Self-evident is the grandeur and magnificence. The 1st floor lays emphasis on banquet hall and swimming pool that has taken as the major function with other sections like dining space, alfresco recreational area for children. Lounge bar has now been a transition linkage as the corridor bridging the lobby and the pool, along which are some modern shapes, like the ceiling decoration and the chandelier that makes an image more casual than solemn.

The pool that deserves more attention for resort facility is particularly embellished with French casement measuring as high as 6 to 7 meters in order to join the bar and the dining room. Through the glass comes sunlight that's bound to enhance the transparency inside and outside.

As for the public space of the first basement, functionality comes first, where to accommodate the club, the banquet room, the family room and video conference. This is more consistent with the needs of entrepreneurs and family banquets. The second vestibule is of gorgeous marble. Along the welcoming lane stands a sculptural wall, a project completely of marble that exerts an immediate impression of resplendence and magnificence.

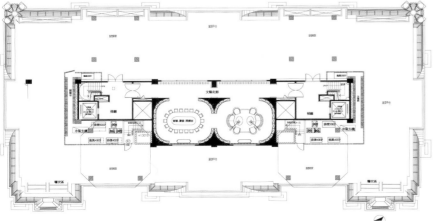
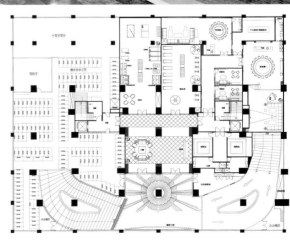

上海绿地海珀风华售楼处
The Sales Center of Sea Ample, Greenland (Shanghai)

设计公司：集艾室内设计（上海）有限公司	Design Company: G & A Design International
设计师：黄全	Designer: Huang Quan
参与设计：方建、赖智	Participant: Fang Jian, Lai Zhi
摄影：三像摄／张静	Photography: Threeimages / Zhang Jing
面积：834 m²	Area: 834 m²

从传统到现代，从东方到西方，跨界的风潮愈演愈烈。2014年底完工的绿地海珀风华售楼处，引入高端奢侈品牌的设计理念，国际化的设计品位，表达新锐的生活态度、审美方式的融合及近距离的奢华体验。

绿地海珀风华位于赵巷别墅区，是绿地集团携手澳大利亚顶尖建筑设计团队精心打造的"海珀"系列高端物业新成员。项目东至置鼎路，西至新通波塘，南至业煌路，北至和尚泾支流，位于总建筑面积约8.9万平方米，规划类独栋别墅为2~3层，叠加别墅5层，容积率1.02，绿化率35%。

绿地海珀风华地理位置优越，西临佘山，东靠虹桥，毗邻沪渝高速赵巷出口，20分钟直通徐家汇、虹桥枢纽等黄金商圈，周边更有宋庆龄国际学校、佘山高尔夫俱乐部、奥特莱斯购物中心、米格天地等国际生活配套。项目是绿地的高端产品系列。

设计师通过后现代手法演绎新古典的表达方式将欧式线条元素精简提取，与不锈钢、镜面、皮质、水晶等现代元素巧妙融合，既有欧式的奢华、气派、高雅的品质感，又不失现代、前沿、新潮的设计感，与项目理念跨界与交融相互呼应，同时也是设计行业的一种全新尝试、突破。每个元素都蕴藏传统与现代，东方与西方，气派与精致的审美品位，也让置身其中的人留下不可磨灭的审美印象。

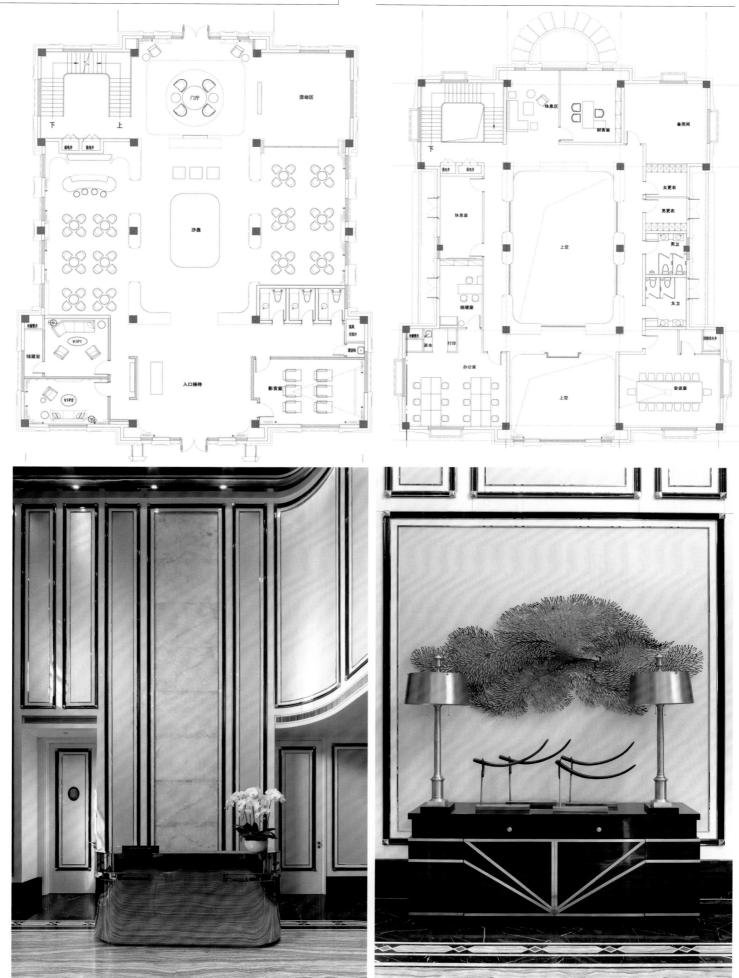

From the tradition to the modern, and from the east to west, the transboundary is being violent. This project finished in the end of 2014 is employed with design philosophy of more high-end luxury land, where to make a luxurious experience in combining a cutting-edge life attitude and aesthetics.

Located in a villa area, this is one of the series high-level real estate by both Greenland and one Austria-based design company. Adjacent to three roads, respectively on its east, west and south and one river on its north, the whole community covers an area of 89 thousand square meters, where to accommodate freestanding villas of 2 or 3 floors, and overlay ones of 5 floors. The plot of ratio reaches 1.02 while its greening rate is 35%.

20 minutes to Xujiahui and Hongqiao transportation hub, the community of Sea Amble not only enjoys a convenient geographical location, neighboring Sheshan Hill on west, Hong qiao on east, and an easy access to Zhao Xiang Slip Road of Shanghai-Chengdu High way. Additionally, it's well facilitated with Soong Ching-ling International School, Sheshan Golf Club, Outlets, and Mega Mills.

With post-modern approaches and neo-classical expressions, quintessence of European lines is fused with modern elements like stainless steel, mirror, leather and crystal. With a luxurious quality sense, momentum and elegance of European style, this space has a design sense that's modern, leading and up-to-date. When corresponding to the cross-field communication, it makes an all-new trial and breakthrough. Each item and every element is both traditional and modern, of the east and the west and has an aesthetic taste of magnificence and delicacy. Here you feel nothing but an indelible aesthetic experience.

合肥海德公馆售楼中心
The Sales Center of Haide Mansion, Hefei

设计公司：上海曼图室内设计有限公司
设计师：冯未墨、张长建、潘超
摄影师：陈志
用材：夹绢玻璃、木饰面、皮革、木纹大理石、灰钛不锈钢
面积：1,200 ㎡

Design Company: Shanghai M2 Design Limited
Designer: Feng Weimo, Zhang Changjian, Pan Chao
Photographer: Chen Zhi
Materials: Wire Glass, Veneer, Leather, Wood-Grain Marble, Gray Tantalized Stainless Steel
Area: 1,200 ㎡

合肥海德公馆售楼中心设计的主基调为现代、自然。建筑入口两层的挑空空间，设计师利用现有空间，以瀑布为灵感，将企业宣传墙以垂壁式的形式伫立在前厅空间，并运用镜面蚀刻和半透的玻璃，营造虚实结合的通高体块，极具视觉冲击力。挑空区起身的灵感来源于雨帘，通过艺术玻璃材质的运用，表达设计师对于现代、生态空间的理解。整体空间以生态原木色为基调，配以局部的皮革和金属、运用灯光的折射，现代感家具与装饰物的组合，共同营造出自然、时尚、现代的体验性空间。

通过销售动线来安排空间布局，注重销售与购买群体的对话关系和尺度感受，在传统人体工学系统下进行家具陈设尺度的微调，来形成舒适的洽谈空间。运用生态的原木、天然岩石样式的陈设品、木纹石材营造生态、自然的整体效果。

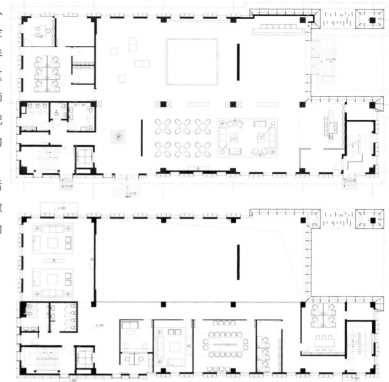

The sales center feels nothing but natural and modern. Its entrance is hollowed-out, measuring as high as two meters. With its space use maximized, waterfall is optimized to erect enterprise propaganda wall in form of vertical type in the antechamber. Meanwhile, the etched glass and transparent mirror altogether bring a real-virtual body to exert a strong visual effect. The inspiration of the hollowed-out space is rooted in rain curtain with artistic glass to embody personal understanding of contemporary and ecological space. Throughout is the keynote by ecological log, with leather and metal in parts, reflective lighting, modern furnishings and accessories to create an experience space that's natural, fashionable and modern.

The sales traffic line is used to guide the spatial arrangement; the dialogue between the sales staff and the customers and the dimension perception are stressed. Furnishings and accessories are of human engineering. So the communication space is no doubt cozy. Of the burly-wood timber, the stone-like accessories, the wood-grain marble and the organic board, all use helps to shape a holistic effect ecological and natural.

肇庆宝能环球金融中心售楼处
The Sales Center of Global Finance, Zhaoqing

设计公司：深圳市昊泽空间设计有限公司
设计师：韩松
用材：灰木纹大理石、黑檀木饰面、黑镜钢、香槟金、皮革
面积：2 800 ㎡

Design Company: Horizon Space Design
Designer: Han Song
Materials: Marble, Ebony Veneer, Black Mirror Steel, Champaign, Leather
Area: 2,800 ㎡

本案位于肇庆开发区，是未来的核心商圈。室内采用后现代设计手法，空间上开阔、通透，隔而不断的围合，灰色基调上跳跃着色彩与高光，随形而生的阵列与发散，强烈对比的视觉冲击，反射与通透得亦真亦幻。流连间、灿烂中，一束追光、一杯红酒、一抹红花……一晃眼，繁华已现。

The location in the Zhaoqing development zone is the business core in the future. Its interior is treated with postmodern approaches. It's wide and broad, transparent, and segmented with all still linked. Against the gray hue are jumping and gloss hues. Arrays are accomplished with the changing shape. The visual effect is strong. When the holistic feels reflective and thorough, and true and illusive, a bunch of light, a red wine, and a touch of red flowers have overspread the prosperity in front of us.

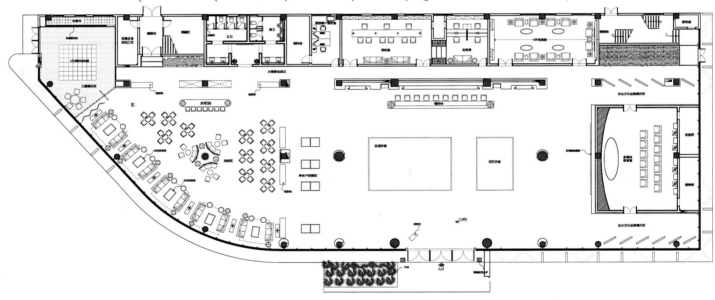

深圳金众·云山栖Hill Villas 售楼中心
The Sales Center of Hill Villas

设计公司：柏年设计
设计师：陈日耿、高洁梅
面积：1 105 m²

Design Company: Hong Kong Century Design Ltd.
Designer: Chen Rigeng, Gao Jiemei
Area: 1,105 m²

金众·云山栖位于大鹏新区行政文化中心核心地段，建筑面积约25万平方米，是大鹏新区规模体量最大的低密度生态宜居大盘。

简欧式的风格，沿袭古典欧式风格的主元素，融入了现代风格的设计手法。在布局上突出轴线对称的仪式感，体现出恢宏的气势，高贵典雅。在细节处理上力求完美，欧式线条的制作工艺精细考究。在色彩的运用上，多以白色、米色为主，深色为辅。相比具有浓厚文化气息的欧式装修风格，简欧更为清新、自然，也符合中国人内敛的审美观念。

以米白色、灰色为主，墨绿色为辅，点缀色为金色。内墙与廊柱采用米白色作为底色，廊柱用墨绿色调剂，整体布局简单明快，高贵中不失典雅，给人一种舒适自在的感觉。金色雍容华贵，华丽中不失庄重。

家具饰品的风格与硬装上的欧式细节是相称的，选择深颜色的欧式座椅沙发，配以同色系的抱枕，金属边大理石材的桌子及接待台，富含意境的抽象挂画，整体空间用暖色调加以协调，构成室内温馨浪漫的气氛。

现有场地华丽大变身，儿童游乐设施搬进售楼处，活跃的颜色、弹性地板、圆角的设施，在这里，让您的孩子释放童真天性，抛却课业压力，尽情游玩。同时这里也是您与孩子沟通情感，享受亲子时刻的最佳去处。

Perched in the core area of the administrative and cultural center, Dapeng New Area, Hill Villas makes the largest low-density residence in the local place with its building area of 25 thousand square meters.

With the main elements of classical European style, the simplified European one has been fused with modern approaches. Its layout stresses the ritual symmetrical sense to embody the grand magnificence. That's noble and refined. Details are deliberately impeccable, like European lines. The dominant white is complimented with beige. Compared with the classical European style, here is more refreshing and natural, more consistent with Chinese reserved aesthetics. The inner wall and the portico are predominated with white and gray, with atrovirens to embellish the portico. All the overall looks crisp and clear, noble and elegant to allow for feelings comfortable and easy. Meanwhile, elegant and poised, the gold is gorgeous and yet reserved.

Furnishing and accessories feature in European style and symmetry, like the dark sofa accompanied with cushions with the same hue, the marble table and the reception desk with metal frame, the connotative wall picture in its abstract sense. The warm color mediates the whole to overspread warm and romantics.

A project this space is which undertakes a gorgeous change. Children recreation facilities have been moved into, which with jumping color, flexible flooring, and the installation with circular beads free children's simplicity and their pressure inflicted by homework, making an ideal place for your and children to communicate.

更正启事

《万有引力 售楼部设计 X》刊登的"江阴尚海荟售楼部"项目介绍中，设计公司和设计师刊登有误，经核实现更正如下：装饰设计公司为桂睿诗设计，设计师为桂峥嵘，本刊编辑部对因错刊给原创设计单位和设计师造成的不良影响深表歉意！

《万有引力 售楼部设计》编辑部

图书在版编目（CIP）数据

万有引力：售楼部设计 XI ／黄滢，马勇 主编 .– 武汉：华中科技大学出版社，2015.9
ISBN 978-7-5680-1273-7
Ⅰ . ①万… Ⅱ . ①黄… ②马… Ⅲ . ①商业建筑—建筑设计—作品集—世界 Ⅳ . ① TU247

中国版本图书馆 CIP 数据核字（2015）第 242448 号

万有引力：售楼部设计 XI

黄滢 马勇 主编

出版发行：华中科技大学出版社（中国·武汉）	
地　　址：武汉市武昌珞喻路 1037 号（邮编：430074）	
出 版 人：阮海洪	
责任编辑：熊纯	责任监印：张贵君
责任校对：岑千秀	装帧设计：筑美文化
印　　刷：中华商务联合印刷（广东）有限公司	
开　　本：965 mm × 1270 mm　1/16	
印　　张：20	
字　　数：160 千字	
版　　次：2016 年 1 月第 1 版 第 1 次印刷	
定　　价：328.00 元（USD 65.99）	

投稿热线：（020）36218949　　　　duanyy@hustp.com
本书若有印装质量问题，请向出版社营销中心调换
全国免费服务热线：400-6679-118 竭诚为您服务
版权所有　侵权必究